Low Reynolds Number Aerodynamics

Low Reynolds Number Aerodynamics

Edited by **Rudd Deakins**

CLANRYE
INTERNATIONAL

New Jersey

Published by Clanrye International,
55 Van Reypen Street,
Jersey City, NJ 07306, USA
www.clanryeinternational.com

Low Reynolds Number Aerodynamics
Edited by Rudd Deakins

International Standard Book Number: 978-1-63240-331-5 (Hardback)

Printed in the United States of America.

Contents

Preface

Over the recent decade, advancements and applications have progressed exponentially. This has led to the increased interest in this field and projects are being conducted to enhance knowledge. The main objective of this book is to present some of the critical challenges and provide insights into possible solutions. This book will answer the varied questions that arise in the field and also provide an increased scope for furthering studies.

This book elucidates information regarding the low Reynolds number aerodynamics. It discusses the recent advancements and trends in the low Re number aerodynamics, transition from laminar to turbulence, unsteady low Reynolds number flows, experimental studies, numerical transition modelling, control of low Re number flows, and MAV wing aerodynamics. It includes contributions by fluid mechanics and aerodynamics scientists and engineers proficient in their respective fields. The studies included in the book demonstrate significant new methods for the realization of the functions of MAV and wind turbine blades.

I hope that this book, with its visionary approach, will be a valuable addition and will promote interest among readers. Each of the authors has provided their extraordinary competence in their specific fields by providing different perspectives as they come from diverse nations and regions. I thank them for their contributions.

Editor

Part 1

Low Reynolds Number Flows

Low Reynolds Number Flows and Transition

M. Serdar Genç[1], İlyas Karasu[1,2], H. Hakan Açıkel[1]
and M. Tuğrul Akpolat[1]
[1]*Wind Engineering and Aerodynamics Research Laboratory, Department of Energy
Systems Engineering, Erciyes University, 38039, Kayseri*
[2]*İskenderun Civil Aviation School, Mustafa Kemal University, 31200, Hatay*
Turkey

1. Introduction

Due to the advances in unmanned aerial vehicles (UAV), micro air vehicles (MAV) and wind turbines, aerodynamics researches concentrated on low Reynolds number aerodynamics, transition and laminar separation bubble (LSB) and its effects on aerodynamic performance. In order to improve endurance, range, efficiency and payload capacity of UAVs, MAVs and wind turbines, the aerodynamic behaviors of these vehicles mentioned should be investigated.

The range of Re numbers of natural and man-made flyers is shown in Figure 1. As the Figure 1 shows most of the commercial and military aircrafts operate on high Reynolds (Re) numbers, and the flow on the surface of these aircraft's wing doesn't separate until the aircraft reaches higher angles of attack -as the angle of attack increases the effects of adverse pressure gradients increase- due to having higher forces of inertia (Genç, 2009). The LSB can be encountered on flyers whose Re number is in the range of 10^4 to 10^6 (King, 2001). On low Re number flow regimes the effects of viscous forces are dominant, which may cause the laminar flow to separate. Under certain circumstances the separated flow which occurs by reason of an adverse pressure gradient reattaches and this forms the LSB. The LSB can be classified as short and long (Tani, 1964). Both short and long bubbles have negative effects on aerodynamic performance. These negative effects may increase drag and decrease lift owing to the altered pressure distribution caused by the presence of the LSB. The characteristics of the LSB depend on the airfoil shape, Re number, surface roughness, freestream disturbances (such as acoustic disturbances), freestream turbulence and geometric discontinuities.

In order to improve the aerodynamic performance, there are new methods being developed to eliminate the effects of the LSB, besides the high lift devices. These methods are called flow control methods and could be classified as active and passive. By using the flow control methods, drag force may be reduced, lift may be increased, stall may be delayed, noise and vibrations may be reduced and reattachment of the separated flow may be obtained.

The effects of the LSB and flow control methods on low Re flow has been investigated by means of various experimental methods, such as force measurement, velocity measurement by using hot-wire anemometry and particle image velocimetry (PIV), pressure measurement with pressure transducers, flow visualization with smoke wire, oil, InfraRed thermography, etc. These systems are useful and accurate but also expensive and everyone cannot find the opportunity to use these methods. Therefore investigating all kind of aerodynamic phenomena via Computational Fluid Dynamics (CFD) is now popular and easier to use. By using CFD, the flow characteristics of a wing profile or the device (UAV, MAV, wind turbine) can be easily analyzed.

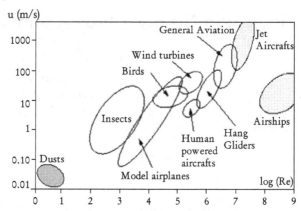

Fig. 1. Flight speeds versus Re number of aircrafts (Chklovski, 2012)

Low Re number flows are seen on mini, micro and unmanned air vehicles, wind turbine blades, model aircrafts, birds and little creatures like bees or flies. Under such low Reynolds numbers, the maximum lift and stall angle are lower than high Re number flow conditions. Owing to the fact that the aerodynamic performance is lower, it is crucial to control of flow and to generate higher lift for this kind of vehicles, devices and/or creatures.

2. Transition

Transition is the phenomenon which occurs in trough different mechanisms in different applications (Langtry & Menter, 2006). The strongest factors affecting transition process are roughness of the wall or surface where the flow passes, adverse pressure gradient and freestream turbulence (Uranga, 2011). Transition is categorized as natural transition, bypass transition, separated flow transition, wake induced transition and reverse transition. There is a parameter to anticipate the type of transition. This parameter is called as acceleration parameter, which represents the effect of freestream acceleration on the boundary layer. The acceleration at the beginning of transition is defined as $K = (v/U^2)(dU/dx)$ (Mayle, 1991). Figure 2 (Mayle, 1991), from which one can decide the type of transition, is plotted as acceleration parameter versus momentum Reynolds number. Above the line marked "*Stability Criterion*" Tollmien-Schlichting type of instability is possible. The separation of a laminar boundary layer occurs above the line marked "*Separation Criterion*". The separation may lead to a separated flow transition. The shaded region on Figure 2 corresponds to the transition Reynolds numbers for turbulence levels between 5% and 10%.

Mayle (1991) presented a study of laminar to turbulent transition phenomena, types of transition and their effects on aerodynamics of gas turbine engines and he also reviewed both theoretical and experimental studies. Schubauer & Skramstad (1947) studied on a flat plate and showed the boundary layer is laminar at local Reynolds numbers (Re_x) lower than 2.8×10^6, whereas the boundary layer is turbulent when Re_x is higher than 2.8×10^6. The boundary layer at Re_x numbers between these two values is called as transitional boundary layer. Formation and type of transition depend on airfoil shape, angle of attack, Re number, free stream turbulence intensity, suction or blowing, acoustic excitation, heating or cooling (White, 1991).

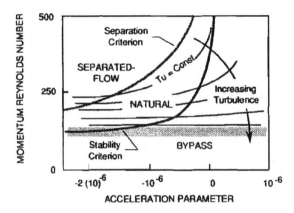

Fig. 2. Topology of the different types of transition in a Reynolds number-acceleration parameter plane (Mayle, 1991)

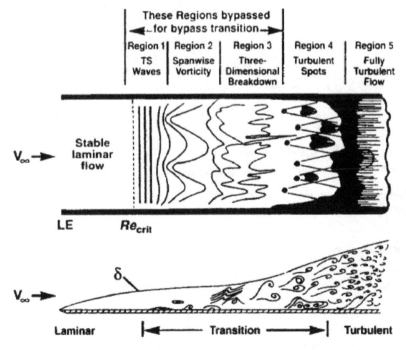

Fig. 3. The natural transition process (Schlichting, 1979)

2.1 Natural transition

This type of transition is seen at high Re numbers and low freestream turbulence levels. Natural transition begins with Tollmien-Schlichting (T/S) waves (Figure 3). T/S waves are the weak instabilities in the laminar boundary layer and this phenomenon was described first by Tollmien and Schlichting (Schlichting, 1979). In order to indicate the T/S waves, a quiet and a relatively less vibrant wind-tunnel and/or experimental apparatus must be employed, based on the fact that the T/S waves are weak instabilities and can be scattered at the higher freestream turbulence levels so freestream turbulence level must be low (<1% (Mayle, 1991)) to observe the T/S waves. Viscosity destabilizes the T/S waves and the waves start to grow very slowly (Langtry & Menter, 2006). The growth of the weak instabilities mentioned, results in nonlinear three-dimensional disturbances. After this certain point the three-dimensional disturbances transform into turbulent spots (Figure 4). The turbulent spots combine and so transition from laminar to turbulent is completed, from now on the flow is fully turbulent. Emmons (1951) and Emmons & Bryson (1951) stated that the turbulent spots within the boundary layer grew and propagated downstream until the flow was fully turbulent. They also presented a model of growth mechanism of turbulent spots, which indicated the time and location dependent random production of the spots.

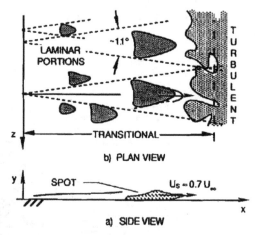

Fig. 4. Turbulent spot geometry and emergence of a turbulent boundary layer trough the growth and propagation of turbulent spots (Mayle, 1991)

2.2 By-pass transition

The other type of transition is bypass transition. As the name suggests, for this type of transition, first, second and third stages of the natural transition process are bypassed (Figure 3). Bypass transition occurs at flows having high freestream turbulence levels. The stages mentioned are bypassed and the turbulent spots are directly produced within the boundary layer by the influence of the freestream disturbances (Mayle, 1991). For bypass transition, linear stability theory is irrelevant and T/S waves have not been documented yet when the freestream turbulence is greater than 1% (Mayle, 1991). So the value 1% can be taken as the boundary between natural and bypass transitions. Lee & Kang (2000) investigated the transition characteristics in a boundary layer over a NACA0012 aerofoil by means of hot-wire anemometry at a range of Reynolds number of $2x10^5$ and $6x10^5$. The aerofoil installed in the incoming wake generated by an aerofoil aligned in tandem with zero angle of attack. The gap between two aerofoils varied from 0.25 to 1.0 of the chord length. Consequently, they pointed that bypass transition occurred in flows around an aerofoil when incoming wave was turbulent and when the incoming wake was present, the transition onset shifted upstream and the transition length became smaller as Re number increased and as the aerofoil gap decreased.

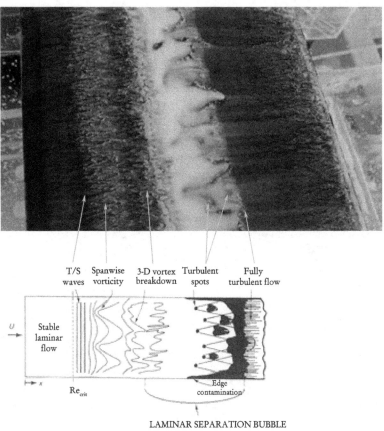

Fig. 5. Comparison of schematic of separation-induced transition process with the experimental photograph obtained oil-flow visualization over the NACA2415 aerofoil (Genç et al., 2012)

2.3 Separated flow transition

At high Re numbers, the laminar boundary layer on an object may transit to turbulent rapidly, and in most cases of high Re number aerodynamics applications, the boundary layer is able to overcome an adverse pressure gradient with minimum disturbance (Tan & Auld, 1992). For low Re number aerodynamics, most of the experimental data indicates the occurrence of flow separation and reattachment in the transitional region (Burgmann et al., 2006; Gaster, 1967; Genç et al., 2008; Genç, 2009; Genç et al., 2011; 2012; Hain et al., 2009; Karasu, 2011; King, 2001; Lang et al., 2004; Mayle, 1991; Mohsen, 2011; Ol et al., 2005; Ricci et al., 2005; Swift, 2009; Tan & Auld, 1992; Tani, 1964; Yang et al., 2007; Yarusevych et al., 2007). The volume full of slowly recirculating air in between the points of separation and reattachment is called *Laminar Separation Bubble* or *Turbulent Reattachment Bubble* (Mayle, 1991).

When a laminar boundary layer cannot overcome the viscous effects and adverse pressure gradients, it separates and transition may occur in the free-shear-layer-like flow near the surface and may reattach to the surface forming a LSB (Mayle, 1991). Flow in the region under the LSB, slowly circulates and reverse flow occurs in this region. The LSB may involve all the stages mentioned for natural transition (Mayle, 1991), but with a LSB stage having the slowly circulating flow region as shown in Figure 5. Genç et al. (2012) carried out experimentally detailed investigation on the LSB over NACA2415 aerofoil by means of oil-flow visualization, pressure measurement and hot-wire anemometry. They compared the flow pattern with the schematic of natural transition introduced by White (White, 1991) and rearranged the figure to adapt the schematic to separated flow transition (Figure 5 and 6).

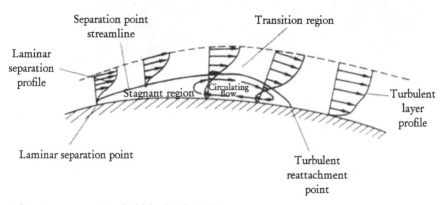

Fig. 6. Laminar separation bubble (Lock, 2007)

Laminar separation bubble may cause adverse effects, such as decreasing of lift force, increasing of drag force, reducing stability of the aircraft, vibration, and noise (Nakano et al., 2007; Ricci et al., 2005; 2007; Zhang et al., 2008). Characteristics of LSB must be understood well to design control system to eliminate to LSB or design new aerofoils which do not affect from adverse effects of LSBs. If Figure 7 (Katz & Plotkin, 1991) is examined carefully, a hump is seen on pressure distribution, this region illuminates the LSB, the region just after the maximum point of this hump indicates transition. If the flow is inviscid, LSB will not take place over the aerofoil.

In a favorable gradient (Figure 8a) the profile is very rounded and there is no point of inflection so separation cannot occur for this case and laminar profiles of this type are very resistant to a transition to turbulence. In a zero pressure gradient (Figure 8b), the point of inflection is at the wall itself. Separation cannot occur here either. The flow will undergo transition at local Reynolds numbers lower than $Re_x = 3x10^6$. In an adverse pressure gradient (Figure

8c to 8e), a point of inflection occurs in the boundary layer. The distance of the point of inflection from the wall increases with the strength of the adverse pressure gradient. For a weak pressure gradient (Figure 8c), flow does not actually separate, but it is vulnerable to transition to turbulence at low Re_x numbers as low as 10^5. For a moderate pressure gradient a critical condition is reached where the wall shear is exactly zero ($\partial u/\partial y=0$). This is defined as the separation point ($\tau_w=0$), because any stronger gradient will actually cause reverse flow at the wall. In this case the boundary layer thickens greatly and the main flow breaks away, or separates from the wall (White, 2004).

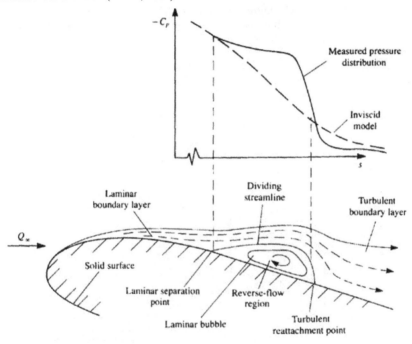

Fig. 7. The effects of laminar separation bubble on pressure distribution (Katz & Plotkin, 1991).

LSBs can be classified as short and long bubbles. The location and size of the bubble is a function of aerofoil shape, angle of attack, freestream disturbances and Re number (Swift, 2009; Tani, 1964). The LSB moves forward and contract in streamwise extent by the increase in angle of attack, which is classified as a short bubble (Tani, 1964). Within this kind of bubble, a small region of constant pressure can be seen, which causes a plateau in the curve of pressure distribution. In consequence of reattachment the curve of the pressure distribution recovers. As the angle of attack increases further, the separation point continues to move towards the leading edge and at a certain angle of attack the flow can no longer reattach to the aerofoil surface within a short distance. This phenomenon is called breakdown or burst of bubble. The occurrence of the breakdown phenomenon does not lead the flow to separate completely. The separated flow passes above the aerofoil and reattaches farther down-stream. The flow region under the separated flow slowly circulates and is called dead-air region or a long bubble. The presence of a short bubble does not significantly alter the peak suction. However, the presence of a long bubble results in a suction plateau of reduced levels in pressure distribution

(Figure 9) over the region occupied by the long bubble and does not result with a sharp suction peak (Tani, 1964). Tan & Auld (1992) experimentally investigated the flow over a Wortmann FX67-150K aerofoil at various Re numbers and various turbulence levels. They concluded that short separation bubbles formed at mild pressure gradient, and that as the pressure gradient increased the short separation bubble burst, forming a long separation bubble. In their experiments, they observed the reattachment of the flow shortly after the transition for the short separation bubble case. But for the long separation bubble case, the separated flow couldn't reattached to the aerofoil surface that quickly. They also stated that if the turbulence level of the freestream increased, length of the bubble decreased because of high energy of the flow, moreover for the short bubble case, the maximum turbulence intensity occurred in the region of reattachment where as the maximum value occurred much forward in the bubble.

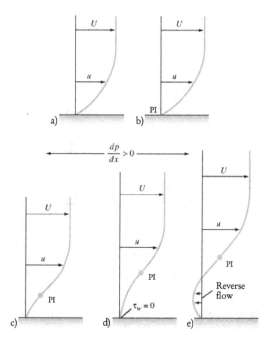

Fig. 8. The effects of various types of pressure gradients on boundary layer (White, 2004)

If the Re number is sufficiently low the separated flow may not reattach to the surface at all, so the laminar separation bubble will not be formed. Therefore the bubble formation is possible only for a certain range of Re numbers. The absence of a bubble at low Re numbers reduces the aerodynamic performance (Tani, 1964). There is a region above the upper surface of the detached flow and near the trailing edge, where the velocity is low and the flow reverses direction in places in a turbulent motion. As the angle of attack increases further, the beginning of the separation moves towards the leading edge of the aerofoil. At a certain angle of attack the lift rapidly falls off as the drag force rapidly increases. This phenomenon is called trailing edge stall. This type of stall is generally encountered on thick aerofoils and often refered as mild stall (McCullough & Gault, 1951). The other type of stall is leading edge stall, and leading edge stall is abrupt (Tani, 1964) laminar flow separation near the leading edge, generally without reattachment and can be encountered for aerofoils with moderate thickness (McCullough & Gault, 1951). For the trailing edge stall the stalled state begins just after the highest lift force obtained. Thin-airfoil stall results from leading edge separation with progressive rearward movement of the point of reattachment. This type of stall occurs on all sharp edged aerofoils and on some thin rounded leading edged aerofoils (McCullough & Gault, 1951).

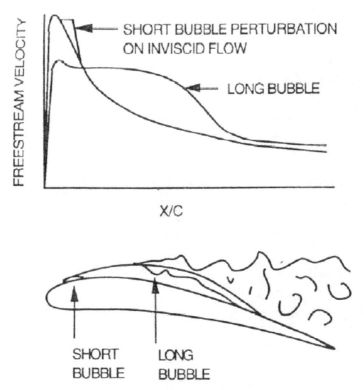

Fig. 9. Separation bubble effects on suction side velocity distribution (Langtry & Menter, 2006)

The effects of different types of stall on the lift coefficient can be seen on Figure 10 (Bak et al., 1998). The angle of attack, Re number, surface roughness and the aerofoil shape influence the stall phenomenon. Yarusecych et al. (2007) investigated NACA0025 aerofoil at a range of Re numbers of 0.55×10^5 to 2.1×10^5 and at three angles of attack ($0°$, $5°$ and $10°$), by means of smoke-wire flow visualization and they observed two boundary layer flow regimes. At $\alpha = 5°$ and Re=0.55×10^5 (Figure 11) the boundary layer on the suction surface of the aerofoil separated and the separated shear layer could not reattach. However, for angle of attack of $5°$ and Re=1.5×10^5 the separated shear layer reattached and this formed a LSB.

(Gaster, 1967) performed an experimental study about LSB by means of constant temperature anemometry (CTA). This study was carried out over a wide range of Re numbers and in a variety of pressure distributions. The bursting circumstances of short bubbles were determined by a unique relationship between Re number and pressure rise. Consequently, LSB was classified as short and long bubble. In the study of Genç et al. (2012), additionally

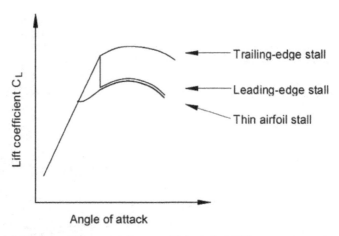

Fig. 10. Sketch of the three different stall types (Bak et al., 1998)

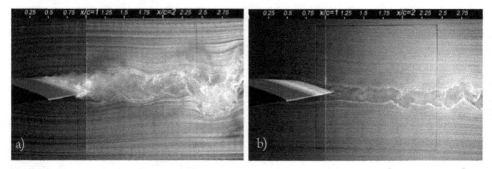

Fig. 11. Flow visualization results for NACA0025 aerofoil at a) Re=0.55x10^5 b) Re=2.1x10^5 (Yarusecych et al., 2007)

long bubble was seen at the angle of attack of 12° for Re=0.5x10^5 (Figure 12), this situation also indicates the bursting of the short bubble at $\alpha = 4°$ and $\alpha = 8°$ when the angle of attack reaches 12°, which leads to forming of a long bubble. The pressure distributions of the other angles of attack (4° and 8°), in which sharp suction peaks can be seen, indicate the presence of the short bubbles. In addition, Figure 12 points out that as the angle of attack increases the LSB moves towards the leading edge. Sharma & Poddar (2010) carried out an experimental study on NACA0015 aerofoil at low Reynolds numbers and at a range of angle of attack (-5° to 25°) and they used the oil flow technique to visualize the transition zone. They obtained the result that as the angle of attack increased the laminar separation bubble moved towards the leading edge and then the bubble burst at a certain angle of attack. The bursting of the bubble caused abrupt stall to occur. Long bubbles should be avoided since they produce large losses and large deviations at higher angles of attack. Short bubbles are effective way of forcing the flow to be turbulent and control the performance. However, one cannot easily predict whether the bubble will be long or short (Mayle, 1991).

Rinioie & Takemura (2004) conducted an experimental study on NACA0012 aerofoil at Re=1.35x10^5 and they concluded that, short bubbles are formed when the angle of attack was less than 11.5°, and long bubbles are formed when the angle of attack is higher than 11.5°.

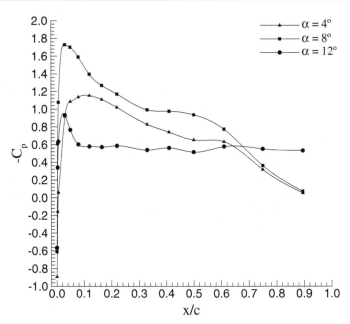

Fig. 12. C_p distributions over the NACA2415 aerofoil at different angles of attack for Re=0.5x10^5 (Genç et al., 2012)

Tan & Auld (1992) experimentally investigated Wortmann FX67-150X aerofoil at low Reynolds numbers and they obtained that as the Reynolds number and freestream turbulence intensity increased, transition occurred earlier and this caused the length of the laminar separation bubble to shorten. Yang et al. (2007) carried out an experimental investigation on GA(W)-1 aerofoil at varying low Re numbers. It was concluded that while the maximum length of the bubble was 20% of the chord length and the maximum height of the bubble was only 1% of the chord length. And also they pointed out that the unsteady vortexes induced by laminar separation bubble were caused by Kelvin-Helmholtz instabilities at angles of attack more than 7°. Diwan & Ramesh (2007) investigated experimentally the length and height of the LSB on a flat plate at different Re numbers. It was obtained that both length and height of the LSB decreased, and reducing ratio of the length is more than that of the height as Re number was risen. Hain et al. (2009) introduced the dynamics of the laminar separation bubbles on low-Reynolds-number aerofoils. It was obtained that Kelvin-Helmholtz instabilities had a weak effect in the spanwise direction and in the later stages of transition these vortices led to a three-dimensional breakdown to turbulence. Lang et al. (2004) also showed that transition in laminar separation bubble was driven by amplification of 2-D T/S waves and first stages of the 3-D disturbances played minor role in transition by studying both experimentally and numerically over an elliptical leading edged flat plate. Furthermore, the results showed that bidirectional vortexes lead to 3-D breakdown. Burgmann et al. (2006) conducted an experimental study on the flow over SD7003 aerofoil which used as wind turbine blades at low Re numbers by means of PIV. They stated that the shear roll-up in the outer region of the LSB causes the regions of concentrated vorticity to form. The vortex roll-up which was initialized by Kelvin-Helmholtz instabilities played effective role at transition process. The results showed that the quasi-periodic development of the large vortex-rolls had a convex or c-like structure (Figure 13). They also mentioned that the c-like structures had no regular pattern in the spanwise direction and that these vortical structures interacted and disturbed each other and most of the vortices maintained their downstream speed, however some vortices decelerated which leaded to vortex-pairing. Their results also indicated that the vortices within the LSB formed as a consequence of the shear layer roll-up due to Kelvin-Helmholtz instabilities and these vortices peeled away from the main recirculation region. These vortices are tend to burst abruptly. The bursting of the vortices causes a strong vertical fluid motion from the wall into the freestream. They also stated that the vortex formed within the LSB increased in size and strength, and its downstream drift speed reduced this low speed state caused instability. The unstable low speed state leaded to a critical condition determined by the accumulation of the vortex strength assumed to be dominated by the momentum ratio and the vortex rotated as a whole structure around the reattachment point in the downstream direction. This leaded to strong ejection of low speed fluid into the freestream. This process acted as a local flow disturbance. The results also showed that the curvature of the aerofoil surface had a distinct effect on the development of the vortices.

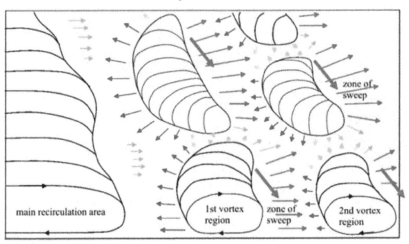

Fig. 13. Sketch of vortex footprint and convex vortex structures (Burgmann et al., 2006)

2.4 Reverse transition

This type of transition is the transition of turbulent to the laminar flow. This is called reverse transition or relaminarization. The relaminarization occurs because of the higher accelerations on the pressure side of most airfoils near the trailing edge, in the exit ducts of combustors and on the suction side of turbine airfoils near the leading edge (Mayle, 1991). Reverse transition is known to occur when the acceleration parameter (K) is greater than 3×10^{-6} (Mayle, 1991) and it is also possible for a relaminarized boundary layer to back to turbulent flow if the acceleration is small enough ($K<3\times10^{-6}$).

2.5 Wake induced transition

Wake induced transition is an instance of the bypass transition which arises in turbomachinery flows where the blade rows subjected to periodically passing turbulent wakes (Langtry &

Menter, 2006). The experimental results showed that the wakes are so disruptive to the laminar boundary layer that turbulent spots often form in the region where the wake is first encountered the aerodynamic body (Langtry & Menter, 2006).

3. Transition modeling

The experimental systems, especially for flow control methods, are expensive and complex. Repeating experiments for a wide range of parameters will naturally cause very expensive solutions. Thus, numerical experimentations using range of CFD methods such as Reynolds Averaged Navier-Stokes (RANS) and/or Direct Numerical and/or Large Eddy Simulation (DNS/LES) methods arise as viable alternatives to experimentation. Furthermore, nowadays with advances in computing technology the CFD methods are used commonly. By using CFD, one can obtain the aerodynamic forces, pressure and velocity distributions over an aerodynamic body and can fix and/or improve the aerodynamic system before the final experimental test (Genç, 2009). Thus, the costs of experimental and design can be decreased.

In parallel with modern developments in experimental capturing, measuring, and identifying the LSBs that are typical for the low-speed flow regimes, improved prediction methods have been devised to account for transition mechanisms over wings of aeroplanes. Today, high performance computing capabilities make it possible to routinely use RANS based CFD methods for simulating low Re number flows. The RANS solvers frequently include practical one- or two-equation turbulence closure models (Wilcox, 1998) for turbulence calculations, although numerical transition modeling side still embed certain difficulties in capturing the complex transition process. Despite the difficulties, transition predictions have developed by means of the e^N method (Cebeci et al., 1972; Drela & Giles, 1987), two-equation low Re-number turbulence models (Cebeci et al., 1972), and some early (Drela & Giles, 1987) and modern (Abu-Ghannam & Shaw, 1994; Wilcox, 1994) methods based on experimental correlations. The e^N method has been quite successful in practice and more or less has become the industry standard (Cebeci et al., 1972). Standard two-equation low-Re models have shown certain successes although the wall damping terms' ability to capture important transition effects limits their use. The correlation-based models (Abu-Ghannam & Shaw, 1994; Suzen & Huang, 2000; 2003) have become helpful tools for industry owing to their use of integral (or global) boundary layer parameters. Recently, transport equation models (Fu & Wang, 2008; Langtry & Menter, 2005; Menter et al., 2004; Walters & Leylek, 2004; 2005) which rely on local information to circumvent some complex procedures in the early methods, have been introduced. These transport equation models solve several transport partial differential equations written for various transition quantities in addition to the baseline turbulence models. Some of these models have been made available in a number of commercial CFD codes such as FLUENT, ANSYS CFX (Langtry & Menter, 2005; Menter et al., 2004). Some of these models are the intermittency transport equation models of Suzen and Huang Suzen & Huang (2000; 2003) and the correlation-based k-ω Shear Stress Transport (SST) transition model of Menter et al. Menter et al. (2004). More recently truly single point RANS approaches such as the k-k_L-ω transition model of Walters and Leylek Walters & Leylek (2004; 2005) which essentially eliminates the need for the intermittency, and a variant of the SST model called as the k-ω-γ model of Fu and Wang Fu & Wang (2008) have been introduced. Such models are suitable for straightforward implementation within RANS methods as they solve additional transport equations for predicting transition phenomena that rely on local information only, in contrast with the global information, as used in the early methods. Assessment of these models has been recently made by different authors including trials of different user-dependent transition correlations (Cutrone et al., 2008; Genç et al., 2008; 2009; Genç, 2009; 2010; Genç et al., 2011; Karasu, 2011; Kaynak & Gürdamar, 2008; Misaka & Obayashi, 2006; Suluksna & Juntasaro, 2008).

Lian & Shyy (2007) conducted a numerical study over a rigid and flexible SD7003 aerofoil and and compared the results with experimental measurements. They investigated the models by coupling a Navier-Stokes solver, the e^N transition model and a Reynolds-averaged two-equation closure to study the laminar separation bubble and transition. Also they proposed a new intermittency function suitable for low Re number transitional flows incurred by laminar separation. They concluded that the LSB became shorter and thinner with the increase of angle of attack. Also higher freestream turbulence levels caused a shorter and thinner separation bubble. And they achieved a good agreement with the model they employed, which is based on linear stability analysis and is designed for steady-state flows with the assumptions that the initial disturbance is small and the boundary layer is thin. Windte et al. (2006) conducted on an experimental and numerical study to investigate LSB over SD 7003 aerofoil at Re=6x10^4. RANS model was used to validation of results of the experiments. They concluded that however both Menter's BSL-2L and Wallin model gave good result, Menter's BSL-2L model gave the best results at both laminar separation and C_L.

The k-ω SST transition model is based on two additional transport equations beyond k and ω: the first is an intermittency equation (γ- equation) that is used to trigger the transition process; and the second is the transition onset momentum thickness Reynolds number ($Re_{\theta t}$-equation) which is forced to follow experimentally-determined correlations with some lag. In this model, SST feature is linked to the transition model by coupling it with the k-ω SST turbulence model (Menter, 1994). Transition correlations are user dependent data retrieved from benchmark experiments obtained at different laboratories. A number of investigators have tried to develop their own correlations of the model parameters against different experimental cases (Cutrone et al., 2008; Fu & Wang, 2008) as the original parameter set remains proprietary (Menter et al., 2004). The k-k_L-ω model is considered as a three-equation eddy-viscosity type, which includes transport equations for turbulent kinetic energy (k), laminar kinetic energy (k_L), and specific dissipation rate (ω). This model, which is essentially a single-point technique, combines the advantages of the prior correlation methods and eliminates the need for intermittency. In this model, the turbulent energy is assumed in the near-wall region to be split into small scale turbulent energy, which contributes directly to turbulence production, and large scale turbulent energy, which contributes to production of laminar kinetic energy through the splat mechanism (Walters & Leylek, 2004; 2005). Walters and Leylek Walters & Leylek (2004; 2005) assumed that transition initiates when the laminar streamwise fluctuations are transported a certain distance from the wall, where that distance is determined by the energy content of the free stream, and the kinematic viscosity. As for the the wall boundary conditions, the k-k_L-ω transition model uses a Neumann type boundary condition which specifies the normal derivative of the function on a surface, whereas the k-ω SST transition model uses Dirichlet type wall boundary conditions which gives the value of the function on a surface (Genç et al., 2011). Recently, Cutrone et al. (2008) proposed to use a combination of the two conditions for ω in the case of separated flows. Catalano & Tognaccini (2010) conducted on a numerical study over SD 7003 aerofoil has Re=6x10^4. In this study, RANS and DNS approach were used. Menter's standart k-ω SST and k-ω SST-LR (Low Reynolds) model were used for RANS approach. Stall and LSB characteristics were predicted same so they concluded k-ω SST-LR could be used for LSB. Sanders et al. (2011) carried out numerical investigation over GH1R low pressure turbine aerofoil using three RANS model and compared with experimental results had been performed before and they concluded that the newer transition model k-k_L-ω model gave better results thank-ω SST and Realizable k-ϵ models according to the experimental values.

4. Experimental techniques at low Reynolds numbers

The wind tunnel tests are crucial for investigations at low Reynolds numbers. In order to understand and improve the performance of low Reynolds number aerofoils, accurate wind tunnel tests must be performed. Since the low Reynolds number aerofoil performance is highly dependent of the laminar boundary layer (Mayle, 1991), low turbulence levels in the wind tunnel's test section are necessary (Selig et al., 2011). If the laminar boundary layer transitions to turbulent prematurely because of freestream disturbances, the phenomenon like laminar separation bubble may not be investigated and/or documented properly. In order to ensure low levels of freestream turbulence and good flow quality in test section of wind tunnel, turbulence screens and honeycombs may be employed. In order to determine aerodynamic characteristics of an aerofoil, wind tunnel test methods such as force and pressure measurements, velocity measurement by using a manometer and pitot static-tube system, hot-wire system and laser doppler anemometry (LDA), laser doppler velocimetry (LDV), and PIV, flow visualizations with oil, smoke wire may be done.

Pressure measurements: Pressure measurement is made by a device at rest relative to the flow. Pressure is usually measured both at walls and in the freestream using the types of measurement device such as pitot static-tube connected to a transducer or manometer. At walls, pressure tappings can be used and can be connected to pressure measurement device via tubes. In order to measure the pressure, one or more transducers can be employed. When a transducer is employed, calibration system requires and calibration can be carried out using by a manometer and pitot-static tube.

Force measurements: During the early years of wind tunnel testing, forces and moments were measured through pan-type *balances*. Although technology has gradually developed, the term *balance* is still used to the devices used for force and moment measurements, today. Balances can be divided into two main groups as internal and external. These names are derived from their location relative to the test model and wind tunnel test section. Internal balances which are almost universally used for measurements in supersonic and transonic tunnels locate inside a model, while external balances which are used in subsonic wind tunnels locate outside the test section of wind tunnel. External balances are with either three or six components. Three-component balances measure lift, drag and pitching moment while six-component balances also measure side force, rolling moment and yawing moment. In external balances, load cell systems are employed. Load cells which simply measures the deformation can be placed on a rod weakened in different axis for different forces and moment. When the wind tunnel is on the weakened part of the rod for the each force will undergo deformation and with the load cells placed on each part one can obtain the data for forces and moment. A balance system software, which is calibrated for forces and moment, gives digital output of forces and moments in desired units. The calibration is performed by loading the load cell with known weights and is repeated before each set of experiments to ensure consistency (Genç et al., 2012).

Velocity measurements: Velocities and turbulence intensities for different points and fields around an object can be measured by using a manometer and pitot-tube system, hot-wire system, and LDV, LDA and PIV. Measurement by using a manometer and pitot-tube system is pressure-based velocity measurement and this method is related to measurement of dynamic pressure. Then, the velocity is calculated with dynamic pressure measured by using Bernoulli equation. This method is simple and inexpensive. The hot-wire system is the most used system, for their very small probes, low response time and high precision for measuring velocity components and turbulence characteristics. This system is capable of detecting turbulent fluctuations with a large dynamic response because of the very small hot-wire thermal inertia and its correction in the anemometer. The hot-wire system can be operated in three methods: constant current (CCA), constant temperature (CTA) and constant voltage (CVA). This systems require also calibration techniques and electronic circuit consisting of a Wheatstone bridge. The probe of a hot-wire system consists of an electrically heated a wire or a thin film. Usually the wire of the probe is made of tungsten or platinum, 0.5-2 mm long and has a diameter of 0.5-5 μm. The films are about 0.1 μm thick and deposited on fine cylinders of quartz, about 25-50 μm in diameter.

Genç et al. (2012) investigated the characteristics of NACA2515 aerofoil at Re numbers of $5x10^4$, $1x10^5$, $2x10^5$ and $3x10^5$ and varying the angle of attack from -12° to 20°. The separation, transition, the formation and progress of the laminar separation bubble and reattached flow were observed obviously by means of force measurements, constant temperature anemometry and pressure measurements. They pointed out that, the near highest point of the peak in the pressure coefficient of suction surface indicated the transition from laminar to turbulent flow and the fluctuations in the graphs of force and moment denoted the separation a post-stall. The C_L-α curves showed that the stall angle and the stall abruptness increased as the Reynolds number raised. Selig et al. (2011) presented a study of a flapped AG455ct aerofoil and a flat-plate with leading edge serration geometries (protuberances like those found on fins/flippers of some aquatic animals) to explore the effects on stall characteristics at low Re numbers and varied angles of attack by means of force measurements. The results for the flapped AG455ct aerofoil showed a dramatic increase with higher flap deflections and the flap efficiency reduces with large deflections up to 40°. And the tests on the flat-plate aerofoil with leading edge serration geometries showed that the serrations on the leading edge lead to higher lift and softer stall and lower drag in the stall and post-stall.

Optical Measurement Techniques: LDV, LDA and PIV techniques are particle-based and optical measurement techniques. These techniques rely on the presence of tracer or seed particles in the flow which not only follow all flow velocity fluctuations but are also sufficient in number to provide the desired spatial or temporal resolution of the measured flow velocity. In these sytems, laser is used to illuminate the desired plane. The laser sheet is placed based on the plane in which velocity will be measured. A combination of cylindrical and spherical lenses is used to adjust both the thickness and the width of the laser sheet. Images are captured using a camera, and a cross-correlation algorithm is used to analyze the images and to calculate the velocities. Ol et al. (2005) compared three different facilities for investigating the LSB on SD7003 aerofoil at low Re number by means of PIV. They conducted experiments in a tow tank, a low-noise wind tunnel and a free-surface water tunnel at Re number of $6x10^4$ and angles of attack of 4°, 8°, 11°. The results showed a qualitative similarity in the bubble shape and velocity fields, as well as the Re stress distributions but the measured location and flow structure of the bubble was still contradictory.

Flow Visualization: Flow visualization is the way to visualize and understand the characterization of the flow. However, air and water which are both used in experiments are transparent, and the observer cannot see the flow and the streamlines around an object with naked eye. So, to make the flow visible, there are two different principles. One is to add different substances into the flow. These substances must be small enough to be able to follow the flow and large enough to be seen. The other principle is to alter the optical properties of the flow. The ratio of refraction of the light passes through the fluid media is a function of the density of the fluid. So within compressible flow, the flow field can be made visible by changing the refractive ratio of light passes through the fluid media (Genç, 2009).

There are different techniques of flow visualization, both optical and by adding different substances. These are smoke visualization, surface flow visualization and optical methods. The smoke visualization method is made by using a smoke-wire or a fog generator. A fog generator usually, generates a single strip of smoke. This may be a disadvantage because, one cannot easily visualize the flow within a large area with just one strip of smoke. The advantage of this method is that the fog generators usually use odorless, non-toxic oil to generate fog. A smoke wire is a high resistant wire or a coil of wires which is stretched between the walls of a wind tunnel and coated with oil. When voltage applied to the smoke wire the wire, gets hot and the oil starts to evaporate to create short bursts of smoke filaments, marking streak lines (Yarusevych et al., 2008a). These filaments introduced to the flow can easily mark the separation and the bubble. To document streak lines, separation and bubble a high-speed camera can be employed for digital imaging (Yarusevych et al., 2008a). As the Reynolds number increases, it gets harder to get a proper image owing to the decrease of smoke filaments' duration. Smoke wire diameter, voltage and the coating liquid employed may be changed as the Reynolds number increases or decreases (Dol et al., 2006). The number of smoke droplets per unit length of the wire depend on the wire diameter and the surface tension of the coating liquid. And also the smoke duration depends on voltage and droplet size (Torii, 1977). This method may not be adequate for higher Reynolds numbers because of being constrained by the smoke duration (Dol et al., 2006; Mueller, 1983; Yarusevych et al., 2008a). (Dol et al., 2006) studied experimentally to determine the optimum smoke-wire material and diameter, wire design (single or coiled), and coating liquid for varying freestream velocities. They concluded that Safex is the most effective liquid and two-coiled Nichrome wire is the optimum wire design to use at freestream velocity of 2 m/s. Yarusevych et al. (2008a) studied experimentally on a NACA0025 aerofoil at low Reynolds numbers by means of smoke-wire visualization and obtained images with a high speed camera. They employed a wire of 0.076 mm diameter and applied 100 volts of voltage to electrically heat and evaporate the coating oil. And they concluded that at $Re_c = 5.5 \times 10^4$ the vortices appeared to form a pattern in the wake region of the aerofoil (approximately, x/c=1 and x/c=2) were similar to a Karman vortex street. Yarusevych et al. (2008b) investigated also the vortex shedding characteristics on a NACA0025 aerofoil at low Reynolds numbers and three different angles of attack by means of constant temperature anemometry and smoke-wire flow visualization method. And they concluded that the shedding frequency of the shear layer roll-up vortices was found to exhibit a power law dependency on the Reynolds number; whereas, the wake vortex shedding frequency varied almost linearly with the Reynolds number. However, the results demonstrated that these correlations depend significantly on the shear layer behavior. Moreover, in contrast to flows over circular cylinders, the ratio of the two frequencies did not exhibit a power law dependency on the Reynolds number.

Yarn tufts (Figure 14) are applied to the surface of the aerofoil and they are used for indicating the flow pattern on the surface of the aerofoil when the wind tunnel is on. They are easily applied to any surface and can be used at any model position. But they do not provide a detailed flow pattern due to moving constantly with the airflow (UWAL, 2012).

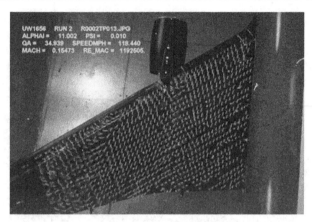

Fig. 14. Fluorescent mini-tufts on an aircraft wing (UWAL, 2012)

Oil-flow surface visualization method is a simple and effective way of documenting the surface flow events by means of photography. In order to photograph the surface flow events using this method, the pigmented oil is applied into a mat black aerofoil and the wind tunnel is on (Genç et al., 2012). Once the oil on the aerofoil's surface is dried, the flow events on the aerofoil surface can be observed and photographed. But it is important to have the proper type of oil mixture for certain wind tunnel speed. The mix should have the right consistency to effectively indicate the development of the boundary layer (Genç et al., 2012). The inertia forces of the moving oil should be lower than the viscous and surface tension forces (Merzkirch, 1974) in order to not affect the flow events on the surface. Some common oils are light diesel oil, light transformer oil and kerosene and some common pigments are titanium dioxide and china clay. Furthermore to see the pigment residue more clearly oleic acid can be added to the mixture (Genç et al., 2012). Genç et al. (2012) investigated experimentally the flow over NACA2415 aerofoil at low Reynolds numbers also by means of oil-flow surface visualization. They applied a mixture using titanium dioxide as pigment, kerosene as oil and oleic acid to see the flow pattern more clearly. They photographed and documented the laminar separation bubble at Reynolds numbers of $1x10^5$, $2x10^5$ and $3x10^5$ and at angles of attack of $4°$, $8°$, $12°$ and $15°$, and compared the results with constant temperature anemometry and pressure measurements experiments. Consequently, they observed the formation and progress of the separation bubble and reattached flow clearly. Selig et al. (2011) studied on E387 aerofoil at low Re numbers by means of oil-flow surface visualization and sketched a graphic of relation between oil-flow visualization photograph and skin friction coefficient (Figure 15). They also mentioned that the texture that existed before running the tunnel still existed in the leading-edge region of the LSB due to the stagnant flow, and that the magnitude of the C_f in this region is quite small because of the low flow speed, and negative in sign because of the reverse flow.

5. Flow control at Low Reynolds numbers

The concept of boundary layer control was introduced first by Prandtl (1904). Flow control methods can be categorized as active and passive flow control methods (Mohsen, 2011; Ricci et al., 2007). Active flow control can be made by adding energy to the free stream or to the boundary layer directly. Passive flow control can be carried out by adding geometrical

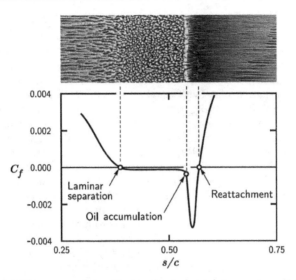

Fig. 15. Conceptual illustration of the relationship between the surface oil flow features and skin friction distribution in the region of a LSB (Selig et al., 2011)

discontinuities or increasing the roughness of the surface. Passive flow control may be simple and cheap but it has its own disadvantages. This kind of geometrical discontinuities increase the drag force and since they are fixed, they cannot adjust with the changing location of the LSB and off-design conditions (Mohsen, 2011; Ricci et al., 2007). Shan et al. (2008) carried out a numerical study on a NACA0012 aerofoil in three different cases. These are uncontrolled flow separation, flow separation control with passive vortex generators and flow separation with control with active vortex generators. And the results showed that in the case of flow separation control with passive vortex generator, the time and spanwise averaged results have shown that the separated flow in the immediate downstream region over an extent of 0.1C. However, the reattached flow separated again and in the conclusion of the transition and reattachment of the separated layer formed the second laminar separation bubble. Thus, the passive vortex generators reduced the size of the separation zone by approximately 80%. The results of the numerical investigation of the active vortex generators case there were no sign of separation so that the flow control with the active vortex generators is more effective than the passive ones. Lengani et al. (2011) investigated the effects of low profile vortex generators on a large-scale flat plate with a prescribed adverse pressure gradient. They placed the vortex generators in the meridional and cross-stream panels and surveyed the velocity fields by means of Laser Doppler Velocimetry (LDV) and measured the total pressure by means of a Kiel total pressure probe. They showed that the presence of vortex generators induced the cross-stream vortices to suppress the separation with large flow oscillations. Santhanakrishnan & Jakob (2005) presented a numerical investigation on a standard Eppler 398 aerofoil with regular perturbations at a range of Re numbers, 2.5×10^4 to 5×10^5. They used smoke-wire flow visualization for qualitative observation of both perturbed and unmodified aerofoils to determine the region of separation. They also employed 2-D PIV measurements to understand the near-wall flow-field behavior. Consequently, at Re=2.5×10^4 and α=4°, separation started very close to the leading edge of the unmodified aerofoil and there was no reattachment. But for the perturbed aerofoil, the flow was attached and the point of the separation was moved further downstream due to the addition of the perturbations. Dassler et al. (2010) developed a new approach for modeling the roughness induced transition, which based on local variables and a transport equation. Two functions determining the value of the transported variable named roughness amplification (Ar), are employed in this model. They implemented the model in the DLR flow solver TRACE and they validated the model by two test cases, a flat plate with roughness and different linear pressure gradients and a flat plate with a two-scale roughness. They indicated the shifting of the transition onset position when different surface roughness values and step change of roughness were prescribed and this result showed that the approach was feasible and was in agreement with the measurements. Roberts & Yaras (1947) conducted an experimental study on a flat surface with five variations. These are three different freestream turbulence intensities (0.5%, 2.5% and 4.5%) and two different Re numbers (3.5×10^5 and 4.7×10^5). They observed both attached flow and separation flow transition with laminar separation bubble. They also mentioned that T/S instability mechanism was responsible for transition in each of the test cases. Consequently, for most of the range of surface roughness heights, the roughness elements remained below the transitioning shear layer of the bubble. This showed that, the roughness elements had no effect on the rate of transition. Ergin & White (2006) carried out an experimental study in a flat plate boundary layer downstream of a spanwise array of cylindrical roughness elements at both subcritical and supercritical values of Re_k. They observed rapid transition only for Re_k=334 because of the sufficiently large fluctuation growth, and they stated that the growth of unsteady disturbance increased with the increasing Re_k. However, for subcritical configurations these disturbances stabilized before the transition could occur. Rizetta & Visbal (2007) used DNS to investigate the effects of roughness elements on a flat plate, for roughness based Re numbers of 202 and 334, and they compared the numerical results with experimental results. The numerical method they employed used a sixth-order-accurate numerical scheme and an overset grid methodology for describing the computational flow-field and a high-order interpolation procedure was employed to maintain accuracy at overlapping boundaries of distinct mesh systems which used for defining the roughness element. For Re_k=202, growth of the integrated turbulent energy was displayed by the simulation in the streamwise extent of computational domain. They also stated that this behavior did not observed experimentally. For Re_k=334, explosive bypass transition displayed by the simulation. Cossu & Brandt (2002; 2004) studied the effect of three-dimensional roughness element in laminar boundary layer. The optimal disturbances in fixed finite magnitude is captured the boundary layer T/S disturbance. They investigated the effects of spanwise invariant disturbances in the T/S unstable frequency band on boundary layer. They found that the stationary finite-amplitude optimal disturbances could suppress the growth of the T/S-like disturbances in a boundary layer.

On the other hand active flow control methods, such as suction/blowing systems, may be expensive and complex but they can adjust with the changing location of the laminar separation bubble by changing the control parameters and/or off-design condition by completely switching of the whole control system. But for active blowing systems some additional disturbances may be generated by the secondary flow through the holes may still be present (Mohsen, 2011). Genç (2009) and Genç et al. (2008; 2009; 2011) studied the prediction of the LSB over the aerofoils at low Re numbers, and the controlling this LSB by using high lift (Genç et al., 2008; 2009; Genç, 2009), blowing and suction systems (Genç, 2009; Genç et al., 2011). The numerical results of the control cases, it was predicted that the separation bubble was eliminated by using the slat, blowing and suction resulting in some marginal increase in the lift and decrease in drag.

Acoustic excitation is an active flow control method, in which typically a signal generator, amplifier and a sound source were used (Ricci et al., 2007). The frequency of the sound wave introduced to the flow can be adjusted with the changing behavior of the aerodynamic body. Theoretically the acoustic excitation induces T/S waves forcing the transition to begin (Ricci et al., 2007). Ricci et al. (2005; 2007) conducted studies on the effects of acoustic disturbances on laminar separation bubbles by means of IR thermography at Reynolds number of $6x10^4$. They inspected a RR3823HL aerofoil at varied angles of attack between $2°$ and $8°$. They introduced a sinusoidal sound wave frequency range was between 200 and 800 Hz with a step of 100 Hz. They concluded that the sinusoidal sound wave having certain frequency reduces the laminar separation bubble's length by retarding the separation. Yarusevych et al. (2007) studied the effect of acoustic excitation amplitude on boundary layer and wake development at low Reynolds numbers by means of hot-wire anemometry, pressure measurements and flow visualization. The results showed that an increase of the excitation amplitude advances the location of reattachment and delays boundary layer separation, reducing the extent of the separation region. Also they indicated that when boundary layer separation occurs without reattachment, the increase of the excitation amplitude above the minimum threshold leads a separation bubble formation with delayed boundary layer separation. Zaman & McKinzie (1991) investigated the effects of acoustic excitation in reducing the adverse influences of the LSB over two dimensional aerofoils at low angles of attack by using smoke wire visualization and hot wire anemometry. They studied in the chord based Re number range of $2.5x10^4 < Re_c < 1x10^5$. However the amplitude of the excitation-induced velocity fluctuation kept constant at a reference point within the flow field, it was founded that the most effective frequency scale was as $U_\infty^{3/2}$. The parameter $St/(Re_c^{1/2}$ corresponding the most effective frequency for all of the cases studied falls in the range of 0.02 to 0.03, with Strouhal number based on the chord. Experimental results showed that lift coefficient had a significant improvement. Zaman (1992) also investigated the effects of acoustic excitation on post-stalled flows over an aerofoil. They used a two dimensional aerofoil LRN (1)-1007 with a chord length of 12.7 cm and employed a crossed hot-film probe for the experiments. The acoustic excitation resulted in a tendency to force the flow to reattach, which was accompanied by an increased lift coefficient and reduced drag coefficient. It was shown that as the amplitude of excitation was increased, a large enhance in the lift was achieved. Ishii (2003) presented the effect of weak acoustic excitation on a separated flow over an aerofoil. Two-dimensional numerical simulations are performed for an NACA0012 aerofoil at angle of attack of $12°$ and Reynolds numbers, $5x10^4$ and $1x10^5$. The amplitude of external sound pressure was set at %0.05 of the static pressure. Numerical results pointed that the acoustic waves with effective frequencies increased the time-averaged lift coefficients. Chang et al. (1992) studied on internal acoustic excitation on the improvement of $NACA63_3 - 068$ aerofoil performance at low Re numbers by means of hot wire and pressure measurements. The acoustic excitation by a loudspeaker was funneled into the interior of the model and then ejected into the flow field from a narrow slot of 0.6 mm in width at %1.25 chord from the leading edge. Experimental results indicated that the flow separation was delayed at the post-stall angle with a low level excitation.

6. Conclusion

In this study, a review of low Reynolds number flows was presented. Firstly, transition and transition types were explained. These transitions types are natural transition, by-pass transition, separated flow transition, reverse transition and wake induced transition. Natural transition is seen at high Reynolds number and low free stream turbulence level. Bypass transition is occurred at high freestream turbulence level and some phases of the natural transition are bypassed. Wake induced transition is an instance of the bypass transition which arises in turbo-machinery flows where the blade rows subjected to periodically passing turbulent wakes. Reverse transition is transition from turbulent flow to laminar. The most important transition for low Reynolds number flows is separated flow transition, in which the flow separates from surface because of viscous effects and adverse pressure gradient, and transition process is completed in the separated region then the fully turbulent flow reattaches to surface. The region between the separation point and the reattachment point is called LSB, which causes negative effects such as decreasing performance, decreasing stability and early stall in aircrafts. LSBs are classified as long bubbles and short bubbles. If short bubble bursts or the separated flow can not reattach to surface and stall will occur and this is a serious problem for aerofoils. Thus, LSB occurring at low Re number flows must be controlled or delayed. Experimental techniques such as pressure measurement, velocity measurement, PIV, smoke and oil flow visualization can be applied for low Reynolds number flows, for instance if pressure distribution is obtained over an aerofoil, a hump is seen at region where laminar separation bubble takes place or another simple method to see LSB region is oil flow visualization method, since there is no movement inside the dead region of LSB, the oil applied before to surface does not move so separation and reattachment point can be seen clearly. Furthermore, transition modeling is one of the most popular research areas nowadays although it has not completely accomplished to model the low Re number flows yet.

7. References

Abu-Ghannam, B.J. & Shaw, R. (1980). Natural Transition of Boundary Layers-The Effect of Turbulence, Pressure Gradient and Flow History. *Journal of Mechanical Engineering Science*, Vol. 22, pp. 213-228.

Bak C., Madsen H.A., Fuglsang P., Rasmussen F. (1998). Double Stall, Risø National Laboratory Technical Report Risø-R-1043(EN), Risø National Laboratory, Roskilde, Denmark.

Burgmann, S., Briicker, C., Shroder, W. (2006). Scanning PIV Measurements of a Laminer Separation Bubble. *Experiments in Fluids*, Vol. 41, pp. 319-326.

Catalano, P. & Tognaccini, R., (2010). Turbulence Modeling for Low-Reynolds-Number Flows. *AIAA Journal*, Vol. 48, pp. 1673-1685.

Cebeci, T., Mosinskis, G.J. and Smith A.M.O. (1972). Calculation of Separation Points in Incompressible Turbulent Flows. *Journal of Aircraft*, Vol. 9, pp. 618-624.

Chang, R.C., Hsiaot, F.B., Shyu, R.N. (1992). Forcing Level Effects of Internal Acoustic Excitation on the Improvement of Airfoil Performance *Journal of Aircraft*, Vol. 29, No. 5, pp. 823-829.

Chklovski, T. (2012) Pointed-tip wings at low Reynolds numbers, University of Southern California, USA, $www - scf.usc.edu/ \sim tchklovs$, Access January 2012.

Cossu, C., Brandt, L.. (2002). Stabilization Tollmien-Schlichting waves by finite amplitude optimal streaks in the Blasius bounday layer. *Physics of Fluids*, Vol. 14, No. 8, pp. L57-L60.

Cossu, C., Brandt, L.. (2004). On of Tollmien-Schlichting waves in streaky boundary layers. *European Journal of Mechanics B/Fluids*, Vol. 23, No. 6, pp. 815-833.

Cutrone, L., De Palma, P., Pascazio, G., Napolitano, M. (2008). Predicting Transition in Two- and Three-dimensional Separated Flows. *International Journal of Heat and Fluid Flow*, Vol. 29, pp. 504-526.

Dassler, P., Kožulovic, D., Fiala, A. (2010). Modelling of Roughness-Induced Transition Using Local Variables. *V European Conference on Computational Fluid Dynamics*, 14-17 June 2010, Lisbon, Portugal.

Diwan, S.S. & Ramesh, O.N. (2007). Laminar separation bubbles: Dynamics and control. *SADHANA-Academy Proceedings In Engineering Science*, Vol. 32, pp. 103-109.

Dol, S.S., Nor, M.A.M., Kamaruzaman, M.K. (2006). An Improved Smoke-Wire Flow Visualization Technique. *Proceedings of the 4th WSEAS International Conference on Fluid Mechanics and Aerodynamics*, Elounda, Greece, August 21-23, pp. 231-236.

Drela, M. & Giles, M.B. (1987). Viscous-inviscid Analysis of Transonic and Low Reynolds Number Aerofoils. *AIAA Journal*, Vol. 25, pp. 1347-1355.

Emmons, H.W. & Bryson, A. (1951). The Laminar-Turbulent Transition in a Boundary Layer-Part II, *Proceeding of 1st U.S. Nat. Congress of Theoratical and Applied Mechanics*, pp. 859-868.

Emmons, H W. (1951). The Laminar-Turbulent Transition in Boundary Layert-Part I. *Journal of the Aeronautical Sciences* , Vol. 18, No. 7, pp. 490-498.

Ergin, F.H. & White, B.E. (2006). Unsteady and transitional flows behind roughness elements. *AIAA Journal*, Vol. 44, No. 11, pp. 2504-2514.

Fu, S. & Wang, L. (2008). Modelling the Flow Transition in Supersonic Boundary Layer with a New k-ω-γ Transition/Turbulence Model. *7th International Symposium on Engineering Turbulence Modelling and Measurements-ETMM7*, Limassol, Cyprus, 4-6 June.

Gaster, M. (1967). The structure and behaviour of separation bubbles. *Aeronautical Research Council Reports and Memoranda*, No:3595, London.

Genç, M.S., Lock, G., Kaynak, Ü. (2008). An experimental and computational study of low Re number transitional flows over an aerofoil with leading edge slat, 8. *AIAA Aviation Technology, Integration and Operations Conference, ATIO 2008*, Anchorage, Alaska, USA.

Genç, M.S., Kaynak, Ü., Lock, G.D. (2009). Flow over an Aerofoil without and with Leading Edge Slat at a Transitional Reynolds Number, *Proc IMechE, Part G- Journal of Aerospace Engineering*, Vol. 223, pp. 217-231.

Genç, M.S. (2009). Control of Low Reynolds Number Flow over Aerofoils and Investigation of Aerodynamic Performance (in Turkish), PhD Thesis, Graduate School of Natural and Applied Sciences, Erciyes University, Kayseri, TURKEY.

Genç, M.S. (2010). Numerical Simulation of Flow over an Thin Aerofoil at High Re Number using a Transition Model, *Proc IMechE, Part C- Journal of Mechanical Engineering Science*, Vol. 224, pp. 2155-2164.

Genç, M.S., Kaynak, Ü., Yapici, H. (2011). Performance of Transition Model for Predicting Low Re Aerofoil Flows without/with Single and Simultaneous Blowing and Suction, *European Journal of Mechanics B/Fluids*, Vol. 30, pp. 218-235.

Genç, M.S., Karasu, İ., Açıkel, H.H. (2012). An experimental study on aerodynamics of NACA2415 aerofoil at low Re numbers, *Experimental Thermal and Fluid Science*, DOI:10.1016/j.expthermflusci.2012.01.029, in press.

Hain, R., Kähler, C., Radespiel, J. (2009). Dynamics of Laminar Separation Bubbles at Low Reynolds Number Aerofoils. *Journal of Fluid Mechanics*, Vol. 630, pp. 129-153.

Ishii, K., Suzuki, S., Adachi, S. (2003). Effect of Weak Sound on Separated Flow over an Airfoil, *Fluid Dynamics Research*, Vol. 33, pp. 357-371.

Karasu, İ. (2011). Experimental and numerical investigations of transition to turbulence and laminar separation bubble over aerofoil at low Reynolds number flows (In Turkish), MSc. Thesis, Graduate School of Natural and Applied Sciences, Erciyes University, Kayseri, TURKEY.

Katz, J. & Plotkin, A. (1991). Low-Speed Aerodynamics from Wing Theory to Panel Methods, McGraw-Hill, Inc.

Kaynak, Ü. and Gürdamar, E. (2008). Boundary-layer Transition under the Effect of Compressibility for the Correlation Based Model", 46th AIAA Aerospace Sciences Meeting, Reno, NV, AIAA-2008-0774, Jan. 07-10.

King R.M. (2001). Study of an Adaptive Mechanical Turbulator for Control of Laminar Separation Bubbles, Degree of Masters of Science Thesis, Graduate Faculty of North Carolina State University, Aerospace Engineering.

Lang, M., Rist, U., Wagner, S. (2004). Investigations on controlled transition development in a laminar separation bubble by means of LDA and PIV. *Experiments in Fluids*, Vol. 36, pp. 43-52.

Langtry, R. & Menter, F. (2005). Transition Modeling for General CFD Applications in Aeronautics, AIAA Paper 2005-0522.

Langtry, R. & Menter, F. (2006). Overview of industrial transition modelling in CFX. ANSYS technical report TPL 8126.

Lee, H. & Kang S.-H. (2000). Flow characteristics of transitional boundary layers on an airfoil in wakes. *Journal of Fluids Engineering*, Vol. 122, No. 3, pp. 522-532.

Lengani, D., Simoni, D., Ubaldi, M., Zunino, P., Bertini, F. (2011). Turbulent boundary layer separation control and loss evaluation of low profile vortex generators, *Experimental Thermal and Fluid Science*, Vol. 35, No. 8, pp. 1505-1513.

Lian, Y. & Shyy W. (2007). Laminar-Turbulent Transition of a Low Reynolds Number Rigid or Flexible Airfoil. *AIAA Journal*, Vol. 45, No.7, pp. 1501-1513.

Lock, G.D. (2007). Lecture notes: Thermofluids 4-fluid mechanics with historical perspective, University of Bath, UK.

Mayle, R.E. (1991). The Role of Laminar-Turbulent Transition in Gas Turbine Engines. *Journal of Turbomachinery*, Vol. 113, 509-537.

McCullough, G.B. & Gault, D.E. (1951). Examples of Three Representative Types of Airfoil-Section Stall At Low Speed. *NACA Technical Note 2502*, Ames Aeronautical Laboratory Moffet Field, California, USA.

Menter, F.R., Langtry, R.B., Likki, S.R., Suzen, Y.B., Huang, P.G., Völker, S. (2004). A Correlation Based Transition Model Using Local Variables: Part I-Model Formulation. *Proceedings of ASME Turbo Expo 2004*, Vienna, Austria, ASME-GT2004-53452, pp. 57-67.

Menter, F. (1994). Two-equation Eddy Viscosity Turbulence Models for Engineering Applications, *AIAA Journal*, Vol. 32, pp. 1598-1605.

Merzkirch, W. (1974). Flow Visualization. London: Academic Press Inc. Ltd.

Misaka, T. & Obayashi, S. (2006). A Correlation-based Transition Models to Flows around Wings. AIAA Paper 2006-918.

Mohsen, J., (2011). Laminar Separation Bubble: Its Structure, Dynamics and control. Chalmers University Of Technology. Research Report 2011:06.

Mueller, T.J. (1983). Flow Visualization by Direct Injection. *Fluid Mechanics Measurements*, Goldstein, R.J., Ed., Hemisphere, Washington, D.C. pp. 307-375.

Nakano, T., Fujisawa, N., Oguma, Y., Takagi, Y., Lee, S. (2007). Experimental study on flow and noise characteristics of NACA0018 airfoil. *Journal of Wind Engineering and Industrial Aerodynamics*, Vol. 95, pp. 511-531.

Ol, M.V., McAuliffe, B.R., Hanff, E.S., Scholz, U., Kähler, C. (2005). Comparison of Laminar Separation Bubble Measurements on a Low Reynolds Number Airfoil in Three Facilities. *35th AIAA Fluid Dynamics Conference and Exhibit*, 6-9 June, Toronto, Ontario Canada, 5149.

Prandtl, L. (1904). On the motion of a fluid with very small viscosity, *Proceedings of 3rd International Mathematics Congress*, Heidelberg, Vol. 3, pp 484-491.

Ricci, R. & Montelpare, S.A. (2005). Quantative IR Thermographic Method to Study the Laminar Separation Bubble Phenomenon. *International Journal of Thermal Sciences*, Vol. 44, pp. 709-719.

Ricci, R., Montelpare, S.A., Silvi, E. (2007). Study of acoustic disturbances effect on laminar separation bubble by IR thermography. *Experimental Thermal and Fluid Science*, Vol. 31, pp. 349-359.

Rinioie, K. & Takemura, N. (2004). Oscillating Behaviour of Laminar Separation Bubble formed on an Aerofoil near Stall, *The Aeronautical Journal*, Vol. 108, pp. 153-163.

Rizetta, D.P., Visbal, M.R., (2007). Direct numerical simulations of flow past an array of distributed roughness elements. *AIAA Journal*, Vol. 45, No. 8, pp. 1967-1975.

Roberts, S.K., Yaras, M.I., (2005). Boundary-Layer Transition Affected by Surface Roughness and Free-Stream Turbulence. *Journal of Fluids Engineering*, Vol. 127, No. 3, pp. 449-457.

Sanders, D.D., O'Brien, W.F., Sondergrad, R., Polanka, M.D., Rabe, D.C. (2011). Prediction Separation and Transitional in Turbine Blades at Low Reynolds Numbers-Part I: Development of Prediction Methodology. *Journal of Turbomachinery*, Vol. 133, pp. 031011/1-031011/9.

Santhanakrishnan, A., Jacob, J.D. (2005). Effect of Regular Surface Perturbations on Flow Over an Airfoil. *35th AIAA Fluid Dynamics Conference and Exhibit*, 6-9 June 2005, Toronto, Ontorio, Canada.

Schlichting, H. (1979). Boundary Layer Theory, McGraw-Hill, Inc, 7th edition.

Schubauer, G.B., Skramstad, H.K. (1947). Laminar boundary layer oscillations and stability of laminar flow. *Journal of Aeronautical Sciences*, Vol. 14, No.2, 69-78.

Selig, M.S., Deters, R. W., Williamson G. A. (2011). Wind Tunnel Testing Airfoils at Low Reynolds Numbers, *49th AIAA Aerospace Sciences Meeting*, 4-7 January, Orlando, FL 875, USA.

Shan, H., Jiang, L., Liu, C., Love, M., Maines, B. (2008). Numerical study of passive and active flow separation control over a NACA0012 airfoil. *Computers & Fluids*, Vol. 37, No. 8, pp. 975-992.

Sharma, M. S. & Poddar, K. (2010). Experimental Investigation of Laminar Separation Bubble for a Flow Past an Airfoil, *Proceedings of ASME Turbo Expo 2010: Power for Land, Sea, and Air (GT2010)*, June 14-18, Glasgow, UK.

Suluksna, K. & Juntasaro, E. (2008). Assessment of Intermittency Transport Equations for Modeling. *International Journal of Heat and Fluid Flow*, Vol. 29, pp. 48-61.

Suzen, Y. B. & Huang, P.G. (2000). Modeling of Flow Transition using an Intermittency Transport Equation. *Journal of Fluids Engineering-Transaction of ASME*, Vol. 122, pp. 273-284.

Suzen, Y. B. & Huang, P.G. (2003). Predictions of Separated and Transitional Boundary Layers under Low-Pressure Turbine Aerofoil Conditions using an Intermittency Transport Equation. *Journal of Turbomachinery*, Vol. 125, pp. 455-464.

Swift, K.M. (2009). An Experimental Analysis of the Laminar Separation Bubble at Low Reynolds Numbers, Msc. Thesis, University of Tennesee Space Institute.

Tan, A.C.N. & Auld, J.D. (1992). Study of Laminar Separation Bubbles at Low Reynolds Number Under Various Conditions. *11th Australasian Fluids Mechanics Conference*, University of Tasmania, Hobart, Australia.

Tani, I. (1964). Low Speed Flows Involving Bubble Separations. *Progress in Aerospace Sciences*, Vol. 5, 70-103.

Torii, K. (1977). Flow Visualization by Smoke-Wire Technique. *Proceedings of the International Symposium on Flow Visualization*, Tokyo, Japan, pp. 251-263.

Uranga, A. (2011). Investigation of transition to turbulence at low Reynolds numbers using Implicit Large Eddy Simulations with a Discontinuous Galerkin method. Ph.D Thesis, Department of Aeronautics and Astronautics, MIT, USA.

University of Washington, Aeronautical Laboratory (UWAL), *http://www.uwal.org/uwalinfo/techguide/flowvis.htm*, access date: January, 2012.

Walters, D.K. & Leylek, J. H. (2004). A New Model for Boundary Layer Transition Using a Single-Point RANS Approach. *Journal of Turbomachinery*, Vol. 126, pp. 193-202.

Walters, D.K. and Leylek, J. H. (2005). Computational fluid dynamics study of wake induced transition on a compressor-like flat plate. *Transactions of the ASME*, Vol. 127, pp. 52-55.

White, F.M. (1991). Viscous fluid flow, Second edition, McGraw-Hill Inc., New York.

White, F.M. (2004). Fluid Mechanics, McGraw-Hill, Inc, 4th edition edition.

Wilcox, D.C. (1994). Simulation of Transition with a Two-equation Turbulence Model. *AIAA Journal*, Vol. 32, pp. 247-255.

Wilcox, D.C. (1998). Turbulence Modeling For CFD, Second Edition, DCW Industries.

Windte J., Scholz, U., Radespiel, R. (2006). Validation of the RANS-simulation of laminar separation bubbles on airfoils. *Aerospace Science and Technology*, Vol. 10, pp. 484-494.

Yang, Z., Haan, L.F., Hui, H. (2007). An Experimental Investigation on the Flow Separation on a Low-Reynolds Number Airfoil. *45th AIAA Aerospace Sciences Meeting and Exhibit*, Reno Nevada.

Yarusevych, S., Kawall, J.G., and Sullivan, P.E. (2007). Separated Shear Layer Transition at Low Reynolds Numbers: Experiments and Stability Analysis, 37th AIAA Fluid Dynamics Conference and Exhibit, 25-28 June, Miami, Florida, USA.

Yarusevych, S., Sullivan, P.E., and Kawall, J.G. (2007). Effect of acoustic excitation amplitude on airfoil boundary layer and wake development. *AIAA Journal*, Vol. 45, No. 4, pp. 760-771.

Yarusevych, S., Sullivan P.E., Kawall, J.G. (2008). Smoke-Wire Flow Visualization on an Airfoil at Low Reynolds Numbers. *38th Fluid Dynamics Conference and Exhibit*, 23-26 June, AIAA-2008-3958, Seattle, Washington, USA.

Yarusevych, S., Kawall, J.G., Sullivan P.E. (2008). Vortex Shedding Characteristics on an Airfoil at Low Reynolds Numbers. *38th Fluid Dynamics Conference and Exhibit*, 23-26 June, AIAA 2008-3957, Seattle, Washington, USA.

Zaman, K.B.M.Q., McKinzie, D.J. (1991). Control of Laminar Separation over Airfoils by Acoustic Excitation. *AIAA Journal*, Vol. 29, pp. 1075-1083.

Zaman K.B. M. Q., (1992). Acoustic Excitation on Stalled Flows over an Airfoil, *AIAA Journal*, Vol. 30, pp. 1492-1498.

Zhang W., Hain R., Kahler C.J. (2008). Scanning PIV Investigation of the Laminar Separation Bubble on a SD7003 Airfoil. *Experiment in Fluids*, Vol. 45, pp. 725-743.

Part 2

Transition Modelling

Transition Modelling for Turbomachinery Flows

F. R. Menter[1] and R. B. Langtry[2]
[1]ANSYS GmbH
[2]The Boeing Company,
[1]Germany
[2]USA

1. Introduction

In the past few decades a significant amount of progress has been made in the development of reliable turbulence models that can accurately simulate a wide range of fully turbulent engineering flows. The efforts by different groups have resulted in a spectrum of models that can be used in many different applications, while balancing the accuracy requirements and the computational resources available to a CFD user. However, the important effect of laminar-turbulent transition is not included in the majority of today's engineering CFD simulations. The reason for this is that transition modelling does not offer the same wide spectrum of CFD-compatible model formulations that is currently available for turbulent flows, even though a large body of publications is available on the subject. There are several reasons for this unsatisfactory situation.

The first is that transition occurs through different mechanisms in different applications. In aerodynamic flows, transition is typically the result of a flow instability (Tollmien-Schlichting waves or in the case of highly swept wings cross-flow instability), where the resulting exponential growth of two-dimensional waves eventually results in a non-linear break-down to turbulence. Transition occurring due to Tollmien-Schlichting waves is often referred to as natural transition [1]. In turbomachinery applications, the main transition mechanism is bypass transition [2] imposed on the boundary layer by high levels of turbulence in the freestream. The high freestream turbulence levels are for instance generated by upstream blade rows. Another important transition mechanism is separation-induced transition [3], where a laminar boundary layer separates under the influence of a pressure gradient and transition develops within the separated shear layer (which may or may not reattach). As well, a turbulent boundary layer can re-laminarize under the influence of a strong favorable pressure gradient [4]. While the importance of transition phenomena for aerodynamic and heat transfer simulations is widely accepted, it is difficult to include all of these effects in a single model.

The second complication arises from the fact that conventional Reynolds averaged Navier-Stokes (RANS) procedures do not lend themselves easily to the description of transitional flows, where both linear and non-linear effects are relevant. RANS averaging eliminates the effects of linear disturbance growth and is therefore difficult to apply to the transition process. While methods based on the stability equations such as the e^n method of Smith & Gamberoni [5] and van Ingen [6] avoids this limitation, they are not compatible with general-purpose CFD methods as typically applied in complex geometries. The reason is that these methods require a priori knowledge of the geometry and the grid topology. In addition, they involve numerous non-local operations (e.g. tracking the disturbance growth along each streamline) that are difficult to implement into today's CFD methods [7]. This is not to argue against the stability approaches, as they are an essential part of the desired "spectrum" of transition models required for the vastly different application areas and accuracy requirements. However, much like in turbulence modeling, it is important to develop engineering models that can be applied in day-to-day operations by design engineers on complicated 3D geometries.

It should be noted that at least for 2D flows, the efforts of various groups has resulted in a number of engineering design tools intended to model transition for very specific applications. The most notable efforts are those of Drela and Giles [8] who developed the XFOIL code which can be used for modeling transition on 2D airfoils and the MISES code of Youngren and Drela [9], which is used for modeling transition on 2D turbomachinery blade rows. Both of these codes use a viscous – inviscid coupling approach which allows the classical boundary layer formulation tools to be used. Transition prediction is accomplished using either an e^n method or an empirical correlation and both of these codes are used widely in their respective design communities. A 3D wing or blade design is performed by stacking the 2D profiles (with the basic assumption that span wise flow is negligible) to create the geometry at which point a 3D CFD analysis is preformed.

Closer inspection shows that hardly any of the current transition models are CFD-compatible. Most formulations suffer from non-local operations that cannot be carried out (with reasonable effort) in general-purpose CFD codes. This is because modern CFD codes use mixed elements and massive parallel execution and do not provide the infrastructure for computing integral boundary layer parameters or allow the integration of quantities along the direction of external streamlines. Even if structured boundary layer grids are used (typically hexahedra), the codes are based on data structures for unstructured meshes. The information on a body-normal grid direction is therefore not easily available. In addition, most industrial CFD simulations are carried out on parallel computers using a domain decomposition methodology. This means in the most general case that boundary layers can be split and computed on different processors, prohibiting any search or integration algorithms. Consequently, the main requirements for a fully CFD-compatible transition model are:

- Allow the *calibrated* prediction of the onset and the length of transition
- Allow the inclusion of different transition mechanisms
- Be formulated locally (no search or line-integration operations)
- Avoid multiple solutions (same solution for initially laminar or turbulent boundary layer)
- Do not affect the underlying turbulence model in fully turbulent regimes
- Allow a robust integration down to the wall with similar convergence as the underlying turbulence model
- Be formulated independent of the coordinate system
- Applicable to three-dimensional boundary layers

Considering the main classes of engineering transition models (stability analysis, correlation based models, low-Re models) one finds that none of these methods can meet all of the above requirements.

The only transition models that have historically been compatible with modern CFD methods are the low-Re models [10,11]. However, they typically suffer from a close interaction with the transition capability and the viscous sublayer modeling and this can prevent an independent calibration of both phenomena [12, 13]. At best, the low-Re models can only be expected to simulate bypass transition which is dominated by diffusion effects from the freestream. This is because the standard low-Re models rely exclusively on the ability of the wall damping terms to capture the effects of transition. Realistically, it would be very surprising if these models that were calibrated for viscous sublayer damping could faithfully reproduce the physics of transitional flows. It should be noted that there are several low-Re models where transition prediction was considered specifically during the model calibration [14, 15, 16]. However, these model formulations still exhibit a close connection between the sublayer behavior and the transition calibration. Re-calibration of one functionality also changes the performance of the other. It is therefore not possible to introduce additional experimental information without a substantial re-formulation of the entire model.

The engineering alternative to low-Re transition models are empirical correlations such as those of [17, 18 and 19]. They typically correlate the transition momentum thickness Reynolds number to local freestream conditions such as the turbulence intensity and pressure gradient. These models are relatively easy to calibrate and are often sufficiently accurate to capture the major effects of transition. In addition, correlations can be developed for the different transition mechanisms, ranging from bypass to natural transition as well as crossflow instability or roughness. The main shortcoming of these models lies in their inherently non-local formulation. They typically require information on the integral thickness of the boundary layer and the state of the flow outside the boundary layer. While these models have been used successfully in special-purpose turbomachinery codes, the non-local operations involved with evaluating the boundary layer momentum thickness and determining the freestream conditions have precluded their implementation into general-purpose CFD codes.

Transition simulations based on linear stability analysis such as the e^n method are the lowest closure level available where the actual instability of the flow is simulated. In the simpler models described above, the physics is introduced through the calibration of the model constants. However, even the e^n method is not free from empiricism. This is because the transition n-factor is not universal and depends on the wind tunnel freestream/acoustic environment and also the smoothness of the test model surface. The main obstacle to the use of the e^n model is that the required infrastructure needed to apply the model is very complicated. The stability analysis is typically based on velocity profiles obtained from highly resolved boundary layer codes that must be coupled to the pressure distribution of a RANS CFD code [7]. The output of the boundary layer method is then transferred to a stability method, which then provides information back to the turbulence model in the RANS solver. The complexity of this set-up is mainly justified for special applications where the flow is designed to remain close to the stability limit for drag reduction, such as laminar wing design.

Large Eddy Simulation (LES) and Direct Numerical Simulations (DNS) are suitable tools for transition prediction [20], although the proper specification of the external disturbance level and structure poses substantial challenges. Unfortunately, these methods are far too costly for engineering applications. They are currently used mainly as research tools and substitutes for controlled experiments.

Despite its complexity, transition should not be viewed as outside the range of RANS methods. In many applications, transition is enforced within a narrow area of the flow due to geometric features (e.g. steps or gaps), pressure gradients and/or flow separation. Even relatively simple models can capture these effects with sufficient engineering accuracy. The challenge to a proper engineering model is therefore mainly in the formulation of a model that can be implemented into a general RANS environment.

In this chapter a novel approach to simulating laminar to turbulent transition is described that can be implemented into a general RANS environment. The central idea behind the new approach is that Van Driest and Blumer's [21] vorticity Reynolds number concept can be used to provide a link between the transition onset Reynolds number from an empirical correlation and the local boundary layer quantities. As a result the model avoids the need to integrate the boundary layer velocity profile in order to determine the onset of transition and this idea was first proposed by [22].

Recently another class of locally formulated transition models have been proposed. They are based on modelling the laminar kinetic energy which is present already upstream of the actual transition location. This information is then applied to trigger the actual transition process. Methods of this kind have been proposed e.g. by Walters and Cokljat [23] and Pacciani et al. [24]. While the argumentation behind the derivation of these models is rather different from the γ-Re_Θ model, the mechanisms by which transition is triggered is very similar.

The current chapter is largely based on Langtry and Menter [25]. More recent articles on model validation and development can be found in [26-28].

2. Model formulation

2.1 Basic concept

The current approach is based on combining experimental correlations with locally formulated transport equations. The essential quantity to trigger the transition process is the vorticity or alternatively the strain rate Reynolds number which is used in the present model is defined as follows:

$$\mathrm{Re}_v = \frac{\rho y^2}{\mu} \left| \frac{\partial u}{\partial y} \right| = \frac{\rho y^2}{\mu} S \tag{1}$$

where y is the distance from the nearest wall, S is the shear strain rate, ρ is the density and μ is the dynamic viscosity. The vorticity Reynolds number it is a local property and can be easily computed at each grid point in an unstructured, parallel Navier-Stokes code.

A scaled profile of the vorticity Reynolds number is shown in Figure 1 for a Blasius boundary layer. The scaling is chosen in order to have a maximum of one inside the boundary layer. This is achieved by dividing the Blasius velocity profile by the corresponding momentum thickness Reynolds number and a constant of 2.193. In other words, the maximum of the profile is proportional to the momentum thickness Reynolds number and can therefore be related to the transition correlations [22] as follows:

$$\text{Re}_\theta = \frac{\max(\text{Re}_v)}{2.193} \tag{2}$$

Based on this observation, a general framework can be built, which can serve as a local environment for correlation based transition models.

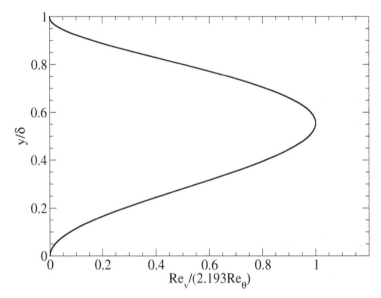

Fig. 1. Scaled vorticity Reynolds number (Re_v) profile in a Blasius boundary layer.

When the laminar boundary layer is subjected to strong pressure gradients, the relationship between momentum thickness and vorticity Reynolds number described by Equation (2) changes due to the change in the shape of the profile. The relative difference between momentum thickness and vorticity Reynolds number, as a function of shape factor (H), is shown in Figure 2. For moderate pressure gradients (2.3 < H< 2.9) the difference between the actual momentum thickness Reynolds number and the maximum of the vorticity Reynolds number is less than 10%. Based on boundary layer analysis a shape factor of 2.3 corresponds to a pressure gradient parameter (λ_θ) of approximately 0.06. Since the majority of experimental data on transition in favorable pressure gradients falls within that range (see for example reference [17]) the relative error between momentum thickness and vorticity Reynolds number is not of great concern under those conditions.

For strong adverse pressure gradients the difference between the momentum thickness and vorticity Reynolds number can become significant, particularly near separation (H = 3.5). However, the trend with experiments is that adverse pressure gradients reduce the transition momentum thickness Reynolds number. In practice, if a constant transition momentum thickness Reynolds number is specified, the transition model is not very

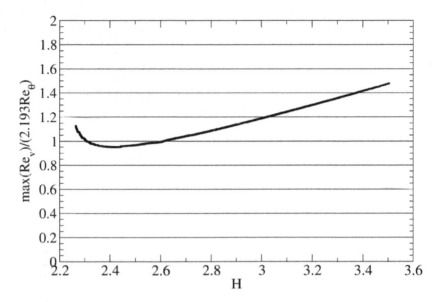

Fig. 2. Relative error between the maximum value of vorticity Reynolds number (Rev) and the momentum thickness Reynolds number (Reθ) as a function of boundary layer shape factor (H).

sensitive to adverse pressure gradients and an empirical correlation such as that of Abu-Ghannam and Shaw [17] is necessary in order to predict adverse pressure gradient transition accurately. In fact, the increase in vorticity Reynolds number with increasing shape factor can actually be used to predict separation induced transition. This is one of the main advantages of the present approach because the standard definition of momentum thickness Reynolds number is not suitable in separated flows.

The function Re_v can be used on physical reasoning, by arguing that the combination of y^2S is responsible for the growth of disturbances inside the boundary layer, whereas $v = \mu / \rho$ is responsible for their damping. As y^2S grows with the thickness of the boundary layer and μ stays constant, transition will take place once a critical value of Re_v is reached. The connection between the growth of disturbances and the function Re_v was shown by Van Driest and Blumer [21] in comparison with experimental data. As well, Langtry and Sjolander [15] found that the location in the boundary layer where Re_v was largest corresponded surprisingly well to the location where the peak growth of disturbances was occurring, at least for bypass transition. The models proposed by Langtry & Sjolander [15] and Walters & Leylek, [16] use Re_v in physics-based arguments based on these observations of disturbance growth in the boundary layer during bypass transition. These models appear superior to conventional low-Re models, as they implicitly contain information of the thickness of the boundary layer. Nevertheless, the close integration of viscous sublayer damping and transition prediction does not easily allow for an independent calibration of both sub-models.

In the present approach first described in references [22, 29, 30 and 31] the main idea is to use a combination of the strain-rate Reynolds number with experimental transition correlations using standard transport equations. Due to the separation of viscous sublayer damping and transition prediction, the new method has provided the flexibility for introducing additional transition effects with relative ease. Currently, the main missing extensions are cross-flow instabilities and high-speed flow correlations and these do not pose any significant obstacles. The concept of linking the transition model with experimental data has proven to be an essential strength of the model and this is difficult to achieve with closures based on a physical modeling of these diverse phenomena.

The present transition model is built on a transport equation for intermittency, which can be used to trigger transition locally. In addition to the transport equation for the intermittency, a second transport equation is solved for the transition onset momentum-thickness Reynolds number. This is required in order to capture the non-local influence of the turbulence intensity, which changes due to the decay of the turbulence kinetic energy in the free-stream, as well as due to changes in the free-stream velocity outside the boundary layer. This second transport equation is an essential part of the model as it ties the empirical correlation to the onset criteria in the intermittency equation. Therefore, it allows the model to be used in general geometries and over multiple airfoils, without additional information on the geometry. The intermittency function is coupled with the SST k-ω based turbulence model [32]. It is used to turn on the production term of the turbulent kinetic energy downstream of the transition point based on the relation between transition momentum-thickness and strain-rate Reynolds number. As the strain-rate Reynolds number is a local property, the present formulation avoids another very severe shortcoming of the correlation-based models, namely their limitation to 2D flows. It therefore allows the simulation of transition in 3D flows originating from different walls. The formulation of the intermittency has also been extended to account for the rapid onset of transition caused by separation of the laminar boundary layer (Equ. 17). In addition the model can be fully calibrated with internal or proprietary transition onset and transition length correlations. The correlations can also be extended to flows with rough walls or to flows with cross-flow instability. It should be stressed that the proposed transport equations do not attempt to model the physics of the transition process (unlike e.g. turbulence models), but form a framework for the implementation of correlation-based models into general-purpose CFD methods. In order to distinguish the present concept from physics based transition modeling, it is named **LCTM** – Local Correlation-based Transition Modeling.

2.2 Transition model equations

The present transition model formulation is described very briefly for completeness, a detailed description of the model and its development can be found in Langtry et al. [25]. It should be noted that a few changes have been made to the model since it was first published [29] in order to improve the predictions of natural transition. These include:

- A new transition onset correlation that results in improved predictions for both natural and bypass transition.
- A modification to the separation induced transition modification that prevents it from causing early transition near the separation point.
- Some adjustments of the model coefficients in order to better account for flow history effects on the transition onset location.

It was expected that different groups will make numerous improvements to the model and consequently a naming convention was introduced in reference [29] in order to keep track of the various model versions. The basic model framework (transport equations without any correlations) was called the γ-Re$_\theta$ transition model. The version number given in reference [29] was called CFX-v-1.0. Based on this naming convention, the present model with the above modifications will be referred to as the γ-Re$_\theta$ model, CFX-v-1.1. The present transition model is briefly summarized in the following pages.

The transport equation for the intermittency, γ, reads:

$$\frac{\partial(\rho\gamma)}{\partial t} + \frac{\partial(\rho U_j\gamma)}{\partial x_j} = P_\gamma - E_\gamma + \frac{\partial}{\partial x_j}\left[\left(\mu + \frac{\mu_t}{\sigma_f}\right)\frac{\partial\gamma}{\partial x_j}\right] \tag{3}$$

The transition sources are defined as follows:

$$P_{\gamma 1} = F_{length}c_{a1}\rho S\left[\gamma F_{onset}\right]^{0.5}\left(1 - c_{e1}\gamma\right) \tag{4}$$

where S is the strain rate magnitude. F_{length} is an empirical correlation that controls the length of the transition region. The destruction/relaminarization source is defined as follows:

$$E_\gamma = c_{a2}\rho\Omega\gamma F_{turb}\left(c_{e2}\gamma - 1\right) \tag{5}$$

where Ω is the vorticity magnitude. The transition onset is controlled by the following functions:

$$Re_V = \frac{\rho y^2 S}{\mu} \tag{6}$$

$$F_{onset1} = \frac{Re_v}{2.193 \cdot Re_{\theta c}} \tag{7}$$

$$F_{onset2} = \min\left(\max\left(F_{onset1}, F_{onset1}^{\;4}\right), 2.0\right) \tag{8}$$

$$R_T = \frac{\rho k}{\mu\omega} \tag{9}$$

$$F_{onset3} = \max\left[1 - \left(\frac{R_T}{2.5}\right)^3, 0\right] \tag{10}$$

$$F_{onset} = \max\left(F_{onset2} - F_{onset3}, 0\right) \tag{11}$$

$Re_{\theta c}$ is the critical Reynolds number where the intermittency first starts to increase in the boundary layer. This occurs upstream of the transition Reynolds number, $\tilde{Re}_{\theta t}$, and the difference between the two must be obtained from an empirical correlation. Both the F_{length} and $Re_{\theta c}$ correlations are functions of $\tilde{Re}_{\theta t}$.

Based on the T3B, T3A, T3A- and the Schubauer and Klebanof test cases a correlation for F_{length} based on $Re_{\theta t}$ from an empirical correlation is defined as:

$$F_{length} = \begin{cases} \left[398.189\cdot10^{-1}+(-119.270\cdot10^{-4})\tilde{R}e_{\theta t}+(-132.567\cdot10^{-6})\tilde{R}e_{\theta t}{}^2\right], \tilde{R}e_{\theta t} < 400 \\ \left[263.404+(-123.939\cdot10^{-2})\tilde{R}e_{\theta t}+(194.548\cdot10^{-5})\tilde{R}e_{\theta t}{}^2+(-101.695\cdot10^{-8})\tilde{R}e_{\theta t}{}^3\right], 400\le\tilde{R}e_{\theta t}<596 \\ \left[0.5-\left(\tilde{R}e_{\theta t}-596.0\right)\cdot3.0\cdot10^{-4}\right], 596\le\tilde{R}e_{\theta t}<1200 \\ \left[0.3188\right], 1200\le\tilde{R}e_{\theta t} \end{cases} \quad (12)$$

In certain cases such as transition at higher Reynolds numbers the transport equation for $\tilde{R}e_{\theta t}$ will often decrease to very small values in the boundary layer shortly after transition. Because F_{length} is based on $\tilde{R}e_{\theta t}$ this can result in a local increase in the source term for the intermittency equation, which in turn can show up as a sharp increase in the skin friction. The skin friction does eventually return back to the fully turbulent value however this effect is unphysical. It appears to be caused by a sharp change in the y^+ in the viscous sublayer where the intermittency decreases back to its minimum value due to the destruction term (Eq. 5). The effect can be eliminated by forcing F_{length} to always be equal to its maximum value (in this case 40.0) in the viscous sublayer. The modification for doing this is shown below. The modification does not appear to have any effect on the predicted transition length. An added benefit is that at higher Reynolds numbers the model now appears to predict the skin friction over shoot measured by experiments.

$$F_{sublayer} = e^{-\left(\frac{R_\omega}{0.4}\right)^2} \quad (13)$$

$$R_\omega = \frac{\rho y^2 \omega}{500\mu} \quad (14)$$

$$F_{length} = F_{length}\left(1-F_{sublayer}\right)+40.0\cdot F_{sublayer} \quad (15)$$

The correlation between $Re_{\theta c}$ and $\tilde{R}e_{\theta t}$ is defined as follows:

$$Re_{\theta c} = \begin{cases} \left[\tilde{R}e_{\theta t}-\begin{pmatrix}396.035\cdot10^{-2}+(-120.656\cdot10^{-4})\tilde{R}e_{\theta t}+(868.230\cdot10^{-6})\tilde{R}e_{\theta t}{}^2 \\ +(-696.506\cdot10^{-9})\tilde{R}e_{\theta t}{}^3+(174.105\cdot10^{-12})\tilde{R}e_{\theta t}{}^4\end{pmatrix}\right], \tilde{R}e_{\theta t}\le1870 \\ \left[\tilde{R}e_{\theta t}-\left(593.11+\left(\tilde{R}e_{\theta t}-1870.0\right)\cdot0.482\right)\right], \tilde{R}e_{\theta t}>1870 \end{cases} \quad (16)$$

The constants for the intermittency equation are:

$$c_{e1}=1.0; \quad c_{a1}=2.0; \qquad c_{e2}=50; \quad c_{a2}=0.06; \quad \sigma_f=1.0;$$

The modification for separation-induced transition is:

$$\gamma_{sep} = \min\left(s_1\max\left[0,\left(\frac{Re_v}{3.235\,Re_{\theta c}}\right)-1\right]F_{reattach}, 2\right)F_{\theta t} \quad (17)$$

$$F_{reattach} = e^{-\left(\frac{R_T}{20}\right)^4} \tag{18}$$

$$\gamma_{eff} = \max\left(\gamma, \gamma_{sep}\right) \tag{19}$$

$$s_1 = 2 \tag{20}$$

The model constants in Equ. 17 have been adjusted from those of Menter et al. [31] in order to improve the predictions of separated flow transition. See Langtry [33] for a detailed discussion of the changes to the model from the Menter et al. [31] version. The main difference is the constant that controls the relation between Re_v and $Re_{\theta c}$ was changed from 2.193, it's value for a Blasius boundary layer, to 3.235, the value at a separation point where the shape factor (H) is 3.5 (see Figure 2). The boundary condition for γ at a wall is zero normal flux while for an inlet γ is equal to 1.0. An inlet γ equal to 1.0 is necessary in order to preserve the original turbulence models freestream turbulence decay rate.

The transport equation for the transition momentum thickness Reynolds number, $\tilde{Re}_{\theta t}$, reads:

$$\frac{\partial\left(\rho \tilde{Re}_{\theta t}\right)}{\partial t} + \frac{\partial\left(\rho U_j \tilde{Re}_{\theta t}\right)}{\partial x_j} = P_{\theta t} + \frac{\partial}{\partial x_j}\left[\sigma_{\theta t}\left(\mu + \mu_t\right)\frac{\partial \tilde{Re}_{\theta t}}{\partial x_j}\right] \tag{21}$$

Outside the boundary layer, the source term $P_{\theta t}$ is designed to force the transported scalar $\tilde{Re}_{\theta t}$ to match the local value of $Re_{\theta t}$ calculated from the empirical correlation (Equ. 35, 36). The source term is defined as follows:

$$P_{\theta t} = c_{\theta t}\frac{\rho}{t}\left(Re_{\theta t} - \tilde{Re}_{\theta t}\right)\left(1.0 - F_{\theta t}\right) \tag{22}$$

$$t = \frac{500\mu}{\rho U^2} \tag{23}$$

where t is a time scale, which is present for dimensional reasons. The time scale was determined based on dimensional analysis with the main criteria being that it had to scale with the convective and diffusive terms in the transport equation. The blending function $F_{\theta t}$ is used to turn off the source term in the boundary layer and allow the transported scalar $\tilde{Re}_{\theta t}$ to diffuse in from the freestream. $F_{\theta t}$ is equal to zero in the freestream and one in the boundary layer. The $F_{\theta t}$ blending function is defined as follows:

$$F_{\theta t} = \min\left(\max\left(F_{wake}\cdot e^{-\left(\frac{y}{\delta}\right)^4}, 1.0 - \left(\frac{\gamma - 1/c_{e2}}{1.0 - 1/c_{e2}}\right)^2\right), 1.0\right) \tag{24}$$

$$\theta_{BL} = \frac{\tilde{Re}_{\theta t}\mu}{\rho U}; \quad \delta_{BL} = \frac{15}{2}\theta_{BL}; \quad \delta = \frac{50\Omega y}{U}\cdot\delta_{BL} \tag{25}$$

$$Re_\omega = \frac{\rho \omega y^2}{\mu}; \quad F_{wake} = e^{-\left(\frac{Re_\omega}{1E+5}\right)^2} \tag{26}$$

The F_{wake} function ensures that the blending function is not active in the wake regions downstream of an airfoil/blade.

The model constants for the $\tilde{Re}_{\theta t}$ equation are:

$$c_{\theta t} = 0.03; \quad \sigma_{\theta t} = 2.0 \tag{27}$$

The boundary condition for $\tilde{Re}_{\theta t}$ at a wall is zero flux. The boundary condition for $\tilde{Re}_{\theta t}$ at an inlet should be calculated from the empirical correlation (Equ. 35, 36) based on the inlet turbulence intensity.

The empirical correlation for transition onset is based on the following parameters:

$$\lambda_\theta = \frac{\rho \theta^2}{\mu} \frac{dU}{ds} \tag{28}$$

$$Tu = 100 \frac{\sqrt{2k/3}}{U} \tag{29}$$

Where dU/ds is the acceleration along the streamwise direction and can be computed by taking the derivative of the velocity (U) in the x, y and z directions and then summing the contribution of these derivatives along the streamwise flow direction:

$$U = \left(u^2 + v^2 + w^2\right)^{\frac{1}{2}} \tag{30}$$

$$\frac{dU}{dx} = \frac{1}{2}\left(u^2 + v^2 + w^2\right)^{-\frac{1}{2}} \cdot \left[2u\frac{du}{dx} + 2v\frac{dv}{dx} + 2w\frac{dw}{dx}\right] \tag{31}$$

$$\frac{dU}{dy} = \frac{1}{2}\left(u^2 + v^2 + w^2\right)^{-\frac{1}{2}} \cdot \left[2u\frac{du}{dy} + 2v\frac{dv}{dy} + 2w\frac{dw}{dy}\right] \tag{32}$$

$$\frac{dU}{dz} = \frac{1}{2}\left(u^2 + v^2 + w^2\right)^{-\frac{1}{2}} \cdot \left[2u\frac{du}{dz} + 2v\frac{dv}{dz} + 2w\frac{dw}{dz}\right] \tag{33}$$

$$\frac{dU}{ds} = \left[(u/U)\frac{dU}{dx} + (v/U)\frac{dU}{dy} + (w/U)\frac{dU}{dz}\right] \tag{34}$$

The use of the streamline direction is not Galilean invariant. However, this deficiency is inherent to all correlation-based models, as their main variable, the turbulence intensity is already based on the local freestream velocity and does therefore violate Galilean invariance. This is not problematic, as the correlations are defined with respect to a wall boundary layer and all velocities are therefore relative to the wall. Nevertheless, multiple moving walls in one domain will likely require additional information.

The use of the streamline direction is not Galilean invariant. However, this deficiency is inherent to all correlation-based models, as their main variable, the turbulence intensity is already based on the local freestream velocity and does therefore violate Galilean invariance. This is not problematic, as the correlations are defined with respect to a wall boundary layer and all velocities are therefore relative to the wall. Nevertheless, multiple moving walls in one domain will likely require additional information.

The empirical correlation has been modified from reference [29] to improve the predictions of natural transition. The predicted transition Reynolds number as a function of turbulence intensity is shown in Figure 3. For pressure gradient flows the model predictions are similar to the Abu-Ghannam and Shaw [17] correlation. The empirical correlation is defined as follows:

$$\mathrm{Re}_{\theta t} = \left[1173.51 - 589.428 Tu + \frac{0.2196}{Tu^2} \right] F(\lambda_\theta), Tu \le 1.3 \tag{35}$$

$$\mathrm{Re}_{\theta t} = 331.50 \left[Tu - 0.5658 \right]^{-0.671} F(\lambda_\theta), Tu > 1.3 \tag{36}$$

$$F(\lambda_\theta) = 1 - \left[-12.986\lambda_\theta - 123.66\lambda_\theta^2 - 405.689\lambda_\theta^3 \right] e^{-\left[\frac{Tu}{1.5} \right]^{1.5}}, \lambda_\theta \le 0 \tag{37}$$

$$F(\lambda_\theta) = 1 + 0.275 \left[1 - e^{[-35.0\lambda_\theta]} \right] e^{\left[\frac{-Tu}{0.5} \right]}, \lambda_\theta > 0 \tag{38}$$

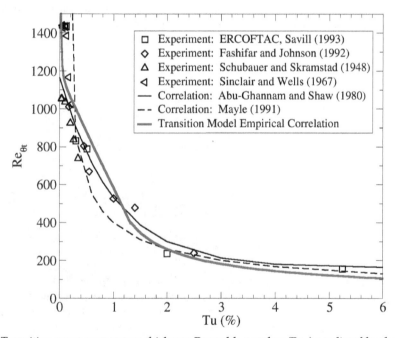

Fig. 3. Transition onset momentum thickness Reynolds number ($\mathrm{Re}_{\theta t}$) predicted by the new correlation as a function of turbulence intensity (Tu) for a flat plate with zero pressure gradient.

For numerical robustness the acceleration parameters, the turbulence intensity and the empirical correlation should be limited as follows:

$$-0.1 \leq \lambda_\theta \leq 0.1$$

$$Tu \geq 0.027$$

$$Re_{\theta t} \geq 20$$

A minimum turbulence intensity of 0.027 percent results in a transition momentum thickness Reynolds number of 1450, which is the largest experimentally observed flat plate transition Reynolds number based on the Sinclair and Wells [36] data. For cases where larger transition Reynolds are believed to occur (e.g. aircraft in flight) this limiter may need to be adjusted downwards.

The empirical correlation is used only in the source term (Eq. 22) of the transport equation for the transition onset momentum thickness Reynolds number. Equations 35 to 38 must be solved iteratively because the momentum thickness (θ_t) is present in the left hand side of the equation and also in the right hand side in the pressure gradient parameter (λ_θ). In the present work an initial guess for the local value of θ_t was obtained based on the zero pressure gradient solution of Eq. 35, 36 and the local values of U, ρ and μ. With this initial guess, equations 35 to 38 were solved by iterating on the value of θ_t and convergence was obtained in less then ten iterations using a shooting point method.

The transition model interacts with the SST turbulence model [32], as follows:

$$\frac{\partial}{\partial t}(\rho k) + \frac{\partial}{\partial x_j}(\rho u_j k) = \tilde{P}_k - \tilde{D}_k + \frac{\partial}{\partial x_j}\left((\mu + \sigma_k \mu_t)\frac{\partial k}{\partial x_j}\right) \tag{39}$$

$$\tilde{P}_k = \gamma_{eff} P_k; \quad \tilde{D}_k = \min\left(\max(\gamma_{eff}, 0.1), 1.0\right) D_k \tag{40}$$

$$R_y = \frac{\rho y \sqrt{k}}{\mu}; \quad F_3 = e^{-\left(\frac{R_y}{120}\right)^8}; \quad F_1 = \max\left(F_{1orig}, F_3\right) \tag{41}$$

where P_k and D_k are the original production and destruction terms for the SST model and F_{1orig} is the original SST blending function. Note that the production term in the ω-equation is not modified. The rationale behind the above model formulation is given in detail in reference [29].

In order to capture the laminar and transitional boundary layers correctly, the grid must have a y^+ of approximately one at the first grid point off the wall. If the y^+ is too large (i.e. > 5) then the transition onset location moves upstream with increasing y^+. All simulations have been performed with CFX-5 using a bounded second order upwind biased discretisation for the mean flow, turbulence and transition equations.

3. Test cases

The remaining part of the chapter will give an overview of some of the public-domain testcases which have been computed with the model described above. This naturally requires a compact representation of the simulations. Most of the cases are described in far more detail in reference [33], including grid refinement and sensitivity studies.

3.1 Flat plate test cases

The flat plate test cases that where used to calibrate the model are the ERCOFTAC T3 series of flat plate experiments [12, 13] and the Schubauer and Klebanof [37] flat plate experiment, all of which are commonly used as benchmarks for transition models. Also included is a test case where the boundary layer experiences a strong favorable pressure gradient that causes it to relaminarize [38]. The inlet conditions for these testcases are summarized in Table 1.

The three cases T3A-, T3A, and T3B have zero pressure gradients with different freestream turbulence intensity (FSTI) levels corresponding to transition in the bypass regime. The Schubauer and Klebanof (S&K) test case has a low free-stream turbulence intensity and corresponds to natural transition. Figure 4 shows the comparison of the model prediction with experimental data for theses cases. It also gives the corresponding FSTI values. In all simulations, the inlet turbulence levels were specified to match the experimental turbulence intensity and its decay rate. This was done by fixing the inlet turbulence intensity and via trial and error adjusting the inlet viscosity ratio (i.e. the ω inlet condition) to match the experimentally measured turbulence levels at various downstream locations. As the freestream turbulence increases, the transition location moves to lower Reynolds numbers.

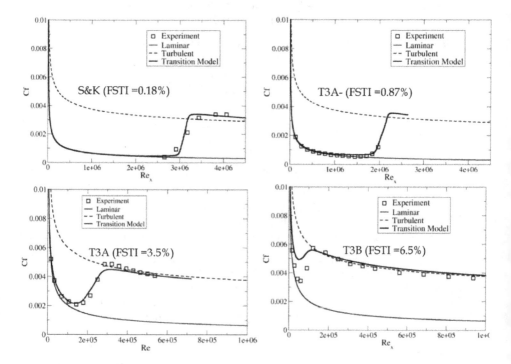

Fig. 4. Results for flat plate test cases with different freestream turbulence levels (FSTI – Freestream Turbulence Intensity).

Case	Inlet Velocity (m/s)	Turbulence Intensity (%) Inlet / Leading Edge value	μ_t / μ	Density (kg/m³)	Dynamic Viscosity (kg/ms)
T3A	5.4	3.3	12.0	1.2	1.8×10^{-5}
T3B	9.4	6.5	100.0	1.2	1.8×10^{-5}
T3A-	19.8	0.874	8.72	1.2	1.8×10^{-5}
Schubauer and Klebanof	50.1	0.3	1.0	1.2	1.8×10^{-5}
T3C2	5.29	3.0	11.0	1.2	1.8×10^{-5}
T3C3	4.0	3.0	6.0	1.2	1.8×10^{-5}
T3C4	1.37	3.0	8.0	1.2	1.8×10^{-5}
T3C5	9.0	4.0	15.0	1.2	1.8×10^{-5}
Relaminarization	1.4	5.5	15	1.2	1.8×10^{-5}

Table 1. Inlet condition for the flat plate test cases.

The T3C test cases consist of a flat plate with a favorable and adverse pressure gradient imposed by the opposite converging/diverging wall. The wind tunnel Reynolds number was varied for the four cases (T3C5, T3C3, T3C2, T3C4) thus moving the transition location from the favorable pressure at the beginning of the plate to the adverse pressure gradient at the end. The cases are used to demonstrate the transition models ability to predict transition under the influence of various pressure gradients. Figure 5 details the results for the

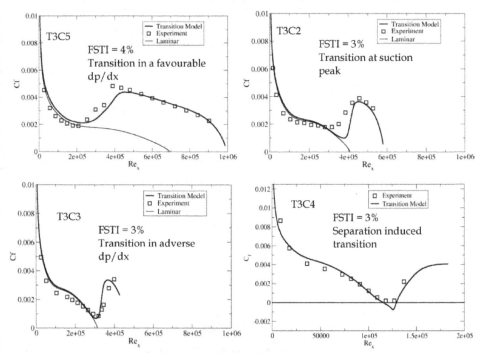

Fig. 5. Results for flat plate test cases where variation of the tunnel Reynolds number causes transition to occur in different pressure gradients (dp/dx).

pressure gradient cases. The effect of the pressure gradient on the transition length is clearly visible with favorable pressure gradients increasing the transition length and adverse pressure gradients reducing it. For the T3C4 case the laminar boundary layer actually separates and undergoes separation induced transition.

The relaminarization test case is shown in Figure 6. For this case the opposite converging wall imposes a strong favorable pressure gradient that can relaminarize a turbulent boundary layer. In both the experiment and in the CFD prediction the boundary layer was tripped near the plate leading edge. In the CFD computation this was accomplished by injecting a small amount of turbulent air into the boundary layer with a turbulence intensity of 3%. The same effect could have been accomplished with a small step or gap in the CFD geometry. Downstream of the trip the boundary layer slowly relaminarizes due to the strong favorable pressure gradient.

For all of the flat plate test cases the agreement with the data is generally good, considering the diverse nature of the physical phenomena computed, ranging from bypass transition to natural transition, separation-induced transition and even relaminarization.

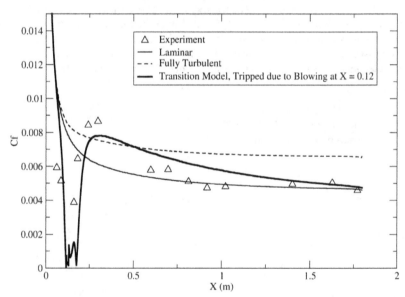

Fig. 6. Predicted skin friction (Cf) for a flat plate with a strong acceleration that causes the boundary layer to relaminarize.

3.2 Turbomachinery test cases

This section descrives a few of the turbomachinery test cases that have been used to validate the transition model including a compressor blade, a low-pressure turbine and a high pressure turbine. A summary of the inlet conditions is shown in Table 2.

For the Zierke and Deutsch [39] compressor blade, transition on the suction side occurs at the leading edge due to a small leading edge separation bubble on the suction side. On the pressure side, transition occurs at about mid-chord. The turbulence contours and the skin

Case	$Re_x = \rho c U_o/\mu$ (x10^6)	Mach $= U_o/a$ where speed of sound (a) $= (\gamma RT)^{0.5}$	Chord (c) (m)	FSTI (%)	μ_t/μ
Zierke and Deutsch Compressor Incidence = -1.5°	0.47	0.1	0.2152	0.18	2.0
Pak-B Low-Pressure Turbine Blade	0.05, 0.075, 0.1	0.03	0.075	0.08, 2.35, 6.0	6.5 - 30
VKI MUR Transonic Guide Vane	0.26	Inlet: 0.15 Outlet: 1.06	0.037	1.0, 6.5	11, 1000

Table 2. Inlet conditions for the turbomachinery test cases.

friction distribution are shown in Figure 7. There appears to be a significant amount of scatter in the experimental data; however, overall the transition model is predicting the major flow features correctly (i.e. fully turbulent suction side, transition at mid-chord on the pressure side). One important issue to note is the effect of stream-wise grid resolution on resolving the leading edge laminar separation and subsequent transition on the suction side. If the number of stream-wise nodes clustered around the leading edge is too low, the model cannot resolve the rapid transition and a laminar boundary layer on the suction side is the result. For the present study, 60 streamwise nodes were used between the leading edge and the $x/C = 0.1$ location.

The Pratt and Whitney PAK-B low pressure turbine blade is a particularly interesting airfoil because it has a loading profile similar to the rotors found in many modern aircraft engines [40]. The low-pressure rotors on modern aircraft engines are extremely challenging flow fields. This is because in many cases the transition occurs in the free shear layer of a separation bubble on the suction side [4]. The onset of transition in the free shear layer determines whether or not the separation bubble will reattach as a turbulent boundary layer and, ultimately, whether or not the blade will stall. The present transition model would therefore be of great interest to turbine designers if it can accurately predict the transition onset location for these types of flows.

Huang et al. [41] conducted experiments on the PAK-B blade cascade for a range of Reynolds numbers and turbulence intensities. The experiments were performed at the design incidence angle for Reynolds numbers of 50,000, 75,000, and 100,000 based on inlet velocity and axial chord length, with turbulence intensities of 0.08%, 2.35% and 6.0% (which corresponded to values of 0.08%, 1.6%, and 2.85% at the leading edge of the blade). The computed pressure coefficient distributions obtained with the transition model and fully turbulent model are compared to the experimental data for the 75 000 Reynolds number, 2.35% turbulent intensity case in Figure 8. On the suction side, a pressure plateau due to a laminar separation with turbulent reattachment exists. The fully turbulent computation completely misses this phenomenon because the boundary layer remains attached over the entire length of the suction surface. The transition model can predict the pressure plateau due to the laminar separation and the subsequent turbulent reattachment location. The pressure side was predicted to be fully attached and laminar.

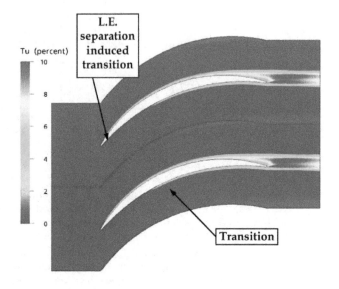

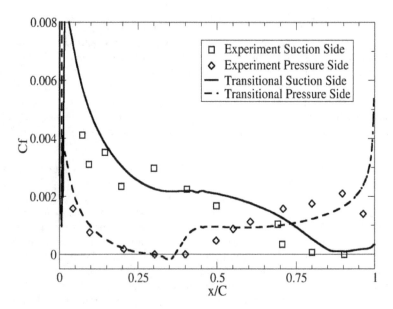

Fig. 7. Turbulence intensity contours (top) and cf-distribution against experimental data (right) for the Zierke & Deutsch compressor.

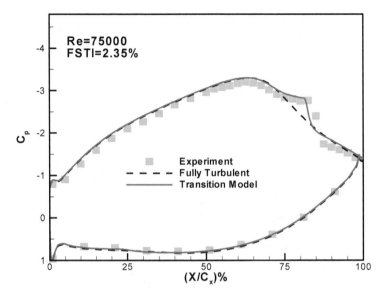

Fig. 8. Predicted blade loading for the Pak-B Low-Pressure turbine at a Reynolds number of 75 000 and a freestream turbulent intensity (FSTI) of 2.35%.

The computed pressure coefficient distributions for various Reynolds numbers and freestream turbulence intensities compared to experimental data are shown in Figure 9. In this figure, the comparisons are organized such that the horizontal axis denotes the Reynolds number whereas the vertical axis corresponds to the freestream turbulence intensity of the specific case. As previously pointed out, the most important feature of this test case is the extent of the separation bubble on the suction side, characterized by the plateau in the pressure distribution. The size of the separation bubble is actually a complex function of the Reynolds number and the freestream turbulence value. As the Reynolds number or freestream turbulence decrease, the size of the separation and hence the pressure plateau increases. The computations with the transition model compare well with the experimental data for all of the cases considered, illustrating the ability of the model to capture the effects of Reynolds number and turbulence intensity variations on the size of a laminar separation bubble and the subsequent turbulent reattachment.

The surface heat transfer for the transonic VKI MUR 241 (FSTI = 6.0%) and MUR 116 (FSTI = 1.0%) test cases [42] is shown in Figure 10. The strong acceleration on the suction side for the MUR 241 case keeps the flow laminar until a weak shock at mid chord, whereas for the MUR 116 case the flow is laminar until right before the trailing edge. Downstream of transition there appears to be a significant difference between the predicted turbulent heat transfer and the measured value. It is possible that this is the result of a Mach number (inlet Mach number $Ma_{inlet}=0.15$, $Ma_{outlet}=1.089$) effect on the transition length [43]. At present, no attempt has been made to account for this effect in the model. It can be incorporated in future correlations, if found consistently important.

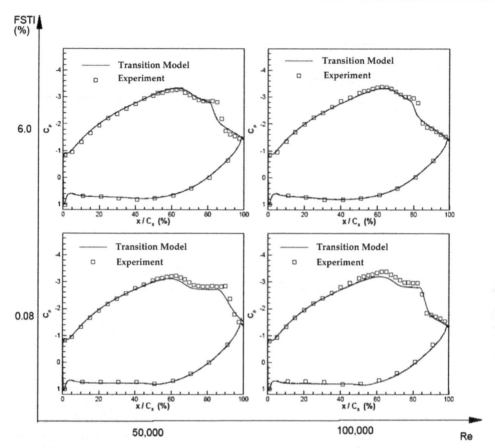

Fig. 9. Blade loading for the Pak-B Low-Pressure turbine at various freestream turbulence intensities (FSTI) and Reynolds numbers (Re).

The pressure side heat transfer is of particular interest for this case. For both cases, transition did not occur on the pressure side, however, the heat transfer was significantly increased for the high turbulence intensity case. This is a result of the large freestream levels of turbulence which diffuse into the laminar boundary layer and increase the heat transfer and skin friction. From a modeling standpoint, the effect was caused by the large freestream viscosity ratio necessary for MUR 241 to keep the turbulence intensity from decaying below 6%, which is the freestream value quoted in the experiment. The enhanced heat transfer on the pressure side was also present in the experiment and the effect appears to be physical. The model can predict this effect, as the intermittency does not multiply the eddy-viscosity but only the production term of the k-equation. The diffusive terms are therefore active in the laminar region.

The S809 airfoil is a 21% thick, laminar-flow airfoil that was designed specifically for horizontal-axis wind turbine (HAWT) applications. The airfoil profile is shown in Figure 11. The experimental results where obtained in the low-turbulence wind tunnel at the Delft

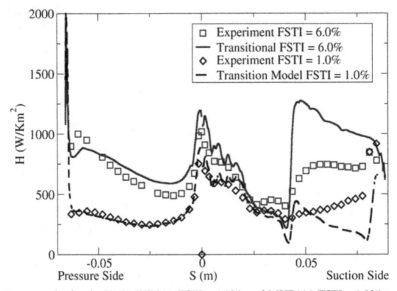

Fig. 10. Heat transfer for the VKI MUR241 (FSTI = 6.0%) and MUR116 (FSTI = 1.0%) test cases.

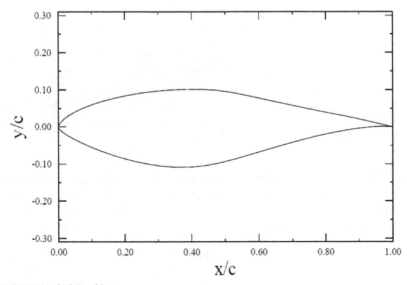

Fig. 11. S809 Airfoil Profile.

University of Technology [44, 45]. The detailed CFD results can be found in reference [46]. The predicted pressure distribution around the airfoil for angles of attack (AoA) of 1° is shown in Figure 12. For the 1° AoA case the flow is laminar for the first 0.5 chord of the airfoil on both the suction and pressure sides. The boundary layers then undergo a laminar separation and reattach as a turbulent boundary layer and this is clearly visible in the experimental pressure distribution plateaus. The fully turbulent computation obviously

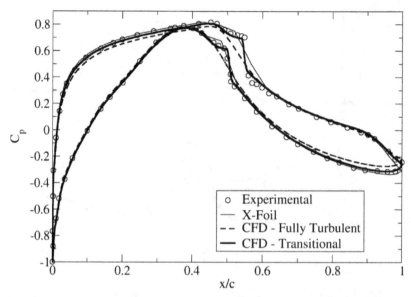

Fig. 12. Pressure distribution (Cp) for the S809 airfoil at 1° angle of attack.

does not capture this phenomenon, as the turbulent boundary layers remain completely attached. Both the transitional CFD and X-Foil solutions do predict the laminar separation bubble. However, X-Foil appears to slightly over predict the reattachment location while the transitional CFD simulation is in very good agreement with the experiment.

Case	Re_x (x10^6)	Mach	Chord (m)	FSTI (%)	μ_t/μ
S809 Airfoil	2.0	0.1	1	0.2	10

Table 3. Inlet conditions for the S809 test case.

The predicted transition locations as a function of angle of attack are shown in Figure 13. The experimental transition locations were obtained using a stethoscope method (Somers, [42]). In general the present transition model would appear to be in somewhat better agreement with the experiment than the X-Foil code, particularly around 14° angle of attack. However, at the moderate angles of attack all of the results appear be to within approximately 5% chord of each other. The X-Foil transition locations appear to change quite rapidly over a few degrees angle of attack while the transition model has a much smoother change in the transition location. The experimental data would appear to confirm that the smooth change in transition location is more physical, however this observation is based primarily on the 10° and 14° angle of attack cases. The results obtained for the lift and drag polars are shown in Figures 14 and 15. Between 0° and 9° the lift coefficients (Cl) predicted by the transitional CFD results are in very good agreement with the experiment while both the XFoil and fully turbulent CFD and results appear to under-predict the lift curve by approximately 0.1. As well, between 0° and 9° the drag coefficient (Cd) predicted by the transitional CFD and X-Foil results are in very good agreement with the experiment while the fully turbulent CFD simulation significantly over predicts the drag, as expected.

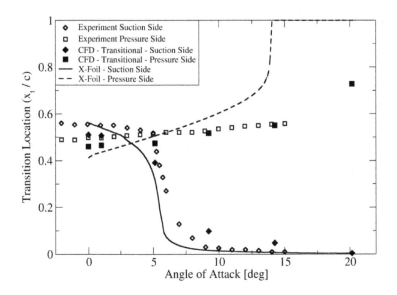

Fig. 13. Transition location (xt/c) vs angle of attack for the S809 airfoil.

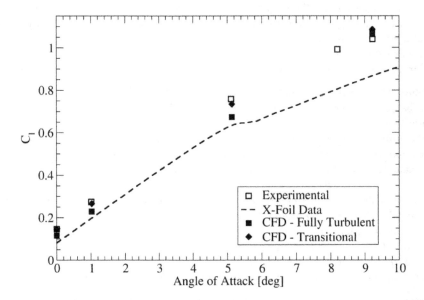

Fig. 14. Lift Coefficient (Cl) Polar for the S809 airfoil.

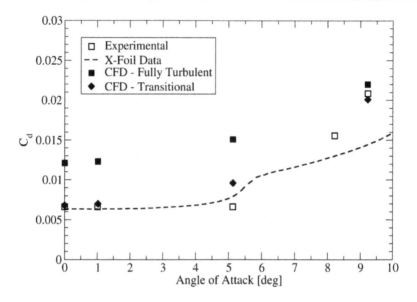

Fig. 15. Drag Coefficient (Cd) Polar for the S809 airfoil The results obtained for the lift and drag polars are.

4. Conclusions

In this chapter various methods for transition prediction in general purpose CFD codes have been discussed. In addition, the requirements that a model has to satisfy to be suitable for implementation into a general purpose CFD code have been listed. The main criterion is that non-local operations must be avoided. A new concept of transition modeling termed Local Correlation-based Transition Model (LCTM) was introduced. It combines the advantages of locally formulated transport equations with the physical information contained in empirical

correlations. The γ-Re$_\theta$ transition model is a representative of that modeling concept. The model is based on two new transport equations (in addition to the k and ω equations), one for intermittency and one for a transition onset criterion in terms of momentum thickness Reynolds number. The proposed transport equations do not attempt to model the physics of the transition process (unlike e.g. turbulence models), but form a framework for the implementation of transition correlations into general-purpose CFD methods.

An overview of the γ-Re$_\theta$ model formulation has been given along with the publication of the full model including some previously undisclosed empirical correlations that control the predicted transition length. The main goal of the present chapter was to publish the full model and release it to the research community so that it can continue to be further validated and possibly extended. Included in this chapter are a number of test cases that can be used to validate the implementation of the model in a given CFD code.

The present transition model accounts for transition due to freestream turbulence intensity, pressure gradients and separation. It is fully CFD-compatible and does not negatively affect the convergence of the solver. Current limitations of the model are that crossflow instability or roughness are not included in the correlations and that the transition correlations are formulated non-Galilean invariant. These limitations are currently being investigated and can be removed in principle

An overview of the test cases computed with the new model has been given. Due to the nature of the chapter, the presentation of each individual test case had to be brief. More details on the test case set-up, boundary conditions grid resolutions etc. can be found in the references. The purpose of the overview was to show that the model can handle a wide variety of geometries and physically diverse problems.

The authors believe that the current model is a significant step forward in engineering transition modeling. Through the use of transport equations instead of search or line integration algorithms, the model formulation offers a flexible environment for engineering transition predictions that is fully compatible with the infrastructure of modern CFD methods. As a result, the model can be used in any general purpose CFD method without special provisions for geometry and grid topology. The authors believe that the LCTM concept of combining transition correlations with locally formulated transport equations has a strong potential for allowing the 1st order effects of transition to be included into today's industrial CFD simulations.

5. Acknowledgments

The model development and validation at ANSYS CFX was funded by GE Aircraft Engines and GE Global Research. The authors would like to thank Dr. Stefan Voelker and Dr. Bill Solomon of GE for their support and numerous thoughtful discussions throughout the course of the model development. As well, Prof. G. Huang of Wright State University and Prof. B. Suzen of North Dakota State University who have supported the original model development with their extensive know-how and their in-house codes. Finally, the authors would also like to thank Dr. Chris Rumsey from the NASA Langley Research Center for supplying the geometry and experimental data for the McDonald Douglas 30P-30N flap and Dr. Helmut Sobieczky of the DLR for his helpful discussions on the DLR F-5 testcase.

6. References

[1] Schlichting, H., 1979, Boundary Layer Theory, McGraw-Hill, 7th edition.
[2] Morkovin, M.V., 1969, "On the Many Faces of Transition", Viscous Drag Reduction, C.S. Wells, ed., Plenum Press, New York, pp 1-31.
[3] Malkiel, E. and Mayle, R.E., 1996, "Transition in a Separation Bubble," ASME Journal of Turbo machinery, Vol. 118, pp. 752-759.
[4] Mayle, R.E., 1991, "The Role of Laminar-Turbulent Transition in Gas Turbine Engines," Journal of Turbomachinery, Vol. 113, pp. 509-537.
[5] Smith, A.M.O. and Gamberoni, N., 1956, "Transition, Pressure Gradient and Stability Theory," Douglas Aircraft Company, Long Beach, Calif. Rep. ES 26388.
[6] Van Ingen, J.L., 1956, "A suggested Semi-Empirical Method for the Calculation of the Boundary Layer Transition Region," Univ. of Delft, Dept. Aerospace Engineering, Delft, The Netherlands, Rep. VTH-74.

[7] Stock, H.W. and Haase, W., 2000, "Navier-Stokes Airfoil Computations with e^N Transition Prediction Including Transitional Flow Regions," AIAA Journal, Vol. 38, No. 11, pp. 2059 – 2066.

[8] Drela, M., and Giles, M. B., 1987, "Viscous-Inviscid Analysis of Transonic and Low Reynolds Number Airfoils", AIAA Journal, Vol. 25, pp. 1347 – 1355.

[9] Youngren, H., and Drela, M., 1991, "Viscous-Inviscid Method for Preliminary Design of Transonic Cascades", AIAA Paper No. 91-2364.

[10] Jones, W. P., and Launder, B. E., 1973, "The Calculation of Low Reynolds Number Phenomena with a Two-Equation Model of Turbulence", Int. J. Heat Mass Transfer, Vol. 15, pp. 301-314.

[11] Rodi, W. and Scheuerer, G., 1984, "Calculation of Laminar-Turbulent Boundary Layer Transition on Turbine Blades", AGARD CP 390 on Heat transfer and colloing in gas turbines.

[12] Savill, A.M., 1993, Some recent progress in the turbulence modeling of by-pass transition, In: R.M.C. So, C.G. Speziale and B.E. Launder, Eds.: Near-Wall Turbulent Flows, Elsevier, p. 829.

[13] Savill, A.M., 1996, One-point closures applied to transition, Turbulence and Transition Modeling, M. Hallbäck et al., eds., Kluwer, pp. 233-268.

[14] Wilcox, D.C.W. (1994). Simulation of transition with a two-equation turbulence model, *AIAA J.* Vol. 32, No. 2.

[15] Langtry, R.B., and Sjolander, S.A., 2002, "Prediction of Transition for Attached and Separated Shear Layers in Turbomachinery", AIAA Paper 2002-3643.

[16] Walters, D.K and Leylek, J.H., 2002, "A New Model for Boundary-Layer Transition Using a Single-Point RANS Approach", ASME IMECE'02, IMECE2002-HT-32740.

[17] Abu-Ghannam, B.J. and Shaw, R., 1980, "Natural Transition of Boundary Layers -The Effects of Turbulence, Pressure Gradient, and Flow History," Journal of Mechanical Engineering Science, Vol. 22, No. 5, pp. 213 – 228.

[18] Mayle, R.E., 1991, "The Role of Laminar-Turbulent Transition in Gas Turbine Engines," Journal of Turbomachinery, Vol. 113, pp. 509-537.

[19] Suzen, Y.B., Huang, P.G., Hultgren, L.S., Ashpis, D.E., 2003, "Predictions of Separated and Transitional Boundary Layers Under Low-Pressure Turbine Airfoil Conditions Using an Intermittency Transport Equation," Journal of Turbomachinery, Vol. 125, No. 3, July 2003, pp. 455-464.

[20] Durbin, P.A., Jacobs, R.G. and Wu, X., 2002, "DNS of Bypass Transition," Closure Strategies for Turbulent and Transitional Flows, edited by B.E. Launder and N.D. Sandham, Cambridge University Press, pp. 449-463.

[21] Van Driest, E.R. and Blumer, C.B., 1963, "Boundary Layer Transition: Freestream Turbulence and Pressure Gradient Effects," AIAA Journal, Vol. 1, No. 6, June 1963, pp. 1303-1306.

[22] Menter, F.R., Esch, T. and Kubacki, S., 2002, "Transition Modelling Based on Local Variables", 5th International Symposium on Engineering Turbulence Modelling and Measurements, Mallorca, Spain.

[23] Walters K. Cokljat, D., (2008), "A Three-Equation Eddy-Viscosity Model for Reynolds-Averaged Navier–Stokes Simulations of Transitional Flow", J. Fluids Eng., Volume 130, Issue 12.

[24] Pacciani R., Marconcini M., Fadai-Ghotbi A., Lardeau S., Leschziner, M. A., 2009, "Calculation of High-Lift Cascades in Low Pressure Turbine Conditions Using a Three-Equation Model" ASME Turbo Expo, Orlando, FL, USA, 8-12 June, ASME paper GT2009-59557. Conf. Proc. Vol. 7: Turbomachinery, Parts A and B, pp. 433-442. ISBN 978-0-7918-4888-3.

[25] Langtry, R.B., Menter, F.R., 2009, "Correlation-Based Transition Modeling for Unstructured Parallelized Computational Fluid Dynamics Codes", AIAA J. Vol. 47, No.12,

[26] Cutrone L., De Palma P., Pascazio G., and Napolitano M., 2008, "Predicting transition in two- and three-dimensional separated flows", Int. J. Heat Fluid Fl 29 504–526.

[27] S. Fu and L. Wang, 2008,: "Modelling the flow transition in supersonic boundary layer with a new k–ω–γ transition/turbulence model", in: 7th International Symposium on Engineering Turbulence Modelling and Measurements-ETMM7, Limassol, Cyprus.

[28] Serdar Genec, M., Kaynak and Ü. Yapıcı, H., 2011,: "Performance of Transition Model for Predicting Low Re Aerofoil Flows without/with Single and Simultaneous Blowing and Suction" European Journal of Mechanics B/Fluids, vol 30 (2), pp. 218-235, 2011.

[29] Menter, F.R., Langtry, R.B., Likki, S.R., Suzen, Y.B., Huang, P.G., and Völker, S., 2006, "A Correlation based Transition Model using Local Variables Part 1- Model Formulation", ASME Journal of Turbomachinery, Vol. 128, Issue 3, pp. 413 – 422.

[30] Langtry, R.B., Menter, F.R., Likki, S.R., Suzen, Y.B., Huang, P.G., and Völker, S., 2006, "A Correlation based Transition Model using Local Variables Part 2 - Test Cases and Industrial Applications", ASME Journal of Turbomachinery, Vol. 128, Issue 3, pp. 423 – 434.

[31] Menter, F.R., Langtry, R.B. and Völker, S., 2006, "Transition Modelling for General Purpose CFD Codes", Journal of Flow, Turbulence and Combustion, Vol. 77, Numbers 1 – 4, pp. 277-303.

[32] Menter, F.R., 1994, "Two-Equation eddy-viscosity turbulence models for engineering applications", AIAA Journal, Vol. 32, No. 8, pp. 1598-1605.

[33] Langtry, R.B., 2006, "A Correlation-Based Transition Model using Local Variables for Unstructured Parallelized CFD codes", Doctoral Thesis, University of Stuttgart. http://elib.uni-stuttgart.de/opus/volltexte/2006/2801/

[34] Fashifar, A. and Johnson, M. W., 1992, "An Improved Boundary Layer Transition Correlation", ASME Paper ASME-92-GT-245.

[35] Schubauer, G.B. and Skramstad, H.K., 1948, "Laminar-boundary-layer oscillations and transition on a flat plate," NACA Rept. 909.

[36] Sinclair, C. and Wells, Jr., 1967, "Effects of Freestream Turbulence on Boundary-Layer Transition," AIAA Journal, Vol. 5, No. 1, pp. 172-174.

[37] Schubauer, G.B. and Klebanoff, P.S., (1955), "Contribution on the Mechanics of Boundary Layer Transition," NACA TN 3489.

[38] McIlroy, H.M. and Budwig, R.S., (2005). "The Boundary Layer over Turbine Blade Models with Realistic Rough Surfaces", *ASME-GT2005-68342, ASME TURBO EXPO 2005*, Reno-Tahoe, Nevada, USA.

[39] Zierke, W.C. and Deutsch, S., 1989, "The measurement of boundary layers on a compressor blade in cascade – Vols. 1 and 2", NASA CR 185118.

[40] Dorney, D.J., Lake, J.P., King, P.L. and Ashpis, D.E., 2000, "Experimental and Numerical Investigation of Losses in Low-Pressure Turbine Blade Rows," AIAA Paper AIAA-2000-0737, Reno, NV.

[41] Huang, J., Corke, T.C., Thomas, F.O., 2003, "Plasma Actuators for Separation Control of Low Pressure Turbine Blades", AIAA Paper No. AIAA-2003-1027.

[42] Arts, T., Lambert de Rouvroit, M., Rutherford, A.W., 1990, "Aero-Thermal Investigation of a Highly Loaded Transonic Linear Turbine Guide Vane Cascade", von Karman Institute for Fluid Dynamics, Technical Note 174.

[43] Steelant, J., and Dick, E., 2001, "Modeling of Laminar-Turbulent Transition for High Freestream Turbulence", Journal of Fluids Engineering, Vol. 123, pp. 22-30.

[44] Somers, D.M., "Design and Experimental Results for the S809 Airfoil", Airfoils, Inc., State College, PA, 1989.

[45] Somers, D.M., "Design and Experimental Results for the S809 Airfoil", NREL/SR-440-6918, January 1997.

[46] Langtry, R.B., Gola, J. and Menter, F.R. 2006, "Predicting 2D Airfoil and 3D Wind Turbine Rotor   Performance using a Transition Model for General CFD Codes", AIAA Paper 2006-0395.

3

Transition at Low-Re Numbers for some Airfoils at High Subsonic Mach Numbers

Ünver Kaynak[1], Samet Çaka Çakmakçıoğlu[1] and Mustafa Serdar Genç[2]
[1]TOBB University of Economics and Technology, Ankara,
[2]Erciyes University, Kayseri
Turkey

1. Introduction

High altitude long endurance UAVs with the role of being low cost satellites due to their long on station times bring renewed interest in Low Reynolds Number Airfoils. These airfoils' behaviors are quite different from their high Reynolds number counterparts [1]. Transition to turbulence and separation bubbles play important roles for these airfoils from low to high Mach numbers. Besides, compressibility makes the stability and transition problems more complex and realistic that are encountered in the transonic regime. As for the density effects, air gets thinner and the Reynolds number starts decreasing as the altitude increases [2]. For instance, at 10,000 m, despite the reduction of gravity of 0.3%, the reduction of air density from 1.225 kg/m^3 to 0.413 kg/m^3 is quite disadvantageous. At 21,000 m, the air density drops significantly to 0.0757 kg/m^3. Meantime, the high altitude UAVs must either fly faster or increase the coefficient of lift to carry the weight. The net result is either increasing the airspeed within the power consumption limits and/or increasing the angle of attack as Re number decreases. Therefore, high altitude flight puts a lot of pressure on designers as to balance the power consumption, high angles of attack nearing stall angles, growing separation bubbles as Re number gets smaller and high subsonic Mach numbers adding the possibility of lambda shocks accelerating the instability in the bubbles. Furthermore, increasing the coefficient of lift moves the cruise point out of the so-called drag bucket limits where the drag increases very quickly. A good selection of a UAV airfoil must account for all these factors for the reason that a good airfoil at sea level may turn out to be a worse selection at high altitude and high alpha conditions.

Predicting low-Reynolds number airfoil performance is a difficult task that requires correctly modeling several flow phenomena such as inviscid flow field with the presence of shock waves, laminar separation regions with presence of separation bubbles, transition to turbulence in the free shear layer and turbulent boundary layer. Especially the presence of the separation bubble may affect the results significantly. Constant pressure assumption across the bounday layer may not be valid across the bubble. Thus, correct modeling of the flow around the airfoils operating at low Reynolds numbers becomes a challenging research problem. Experimental study for low-Reynolds number flows also have some certain difficulties. In a low Reynolds number experimental work regarding the drag coefficient measurements on a Wortmann FX63-137 airfoil at the same test conditions at three different research facilities, it is reported that the results show differences of more than 50% [3].

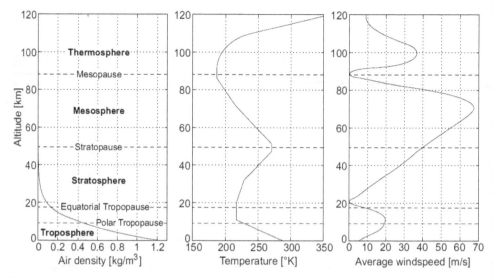

Fig. 1. Variation of density, temperature and wind speed with the altitude [2].

High altitude long endurance vehicles' flight conditions impose difficult testing conditions for the ground based test facilities to emulate the low density, low freestream turbulence and high subsonic Mach numbers at the same time. Flight testing would be required to collect the actual experimental data. For instance, a high altitude sailplane project called APEX was started by NASA that aimed for collecting boundary layer data after being released from a very high altitude of 108 K ft. after a balloon launch at that altitude [4]. However, it was clear for the designers of the aircraft that computational methodology was very much required to improve our know-how on laminar and transitional boundary layers at high Mach numbers. In reference [4], a viscous-inviscid interaction methodology [5,6] and time-accurate RANS [7,8] were used in the numerical predictions. Today, state of the art Reynolds Averaged Navier-Stokes (RANS) solvers are widely available for numerically predicting fully turbulent part of flow fields, but none of these models are adequate to handle flows with significant transition effects because of lack of practical transition modeling. Nevertheless, transition predictions have shown certain progress and utility by means of the well-known e^N method [5], some two-equation low Re-number turbulence models [9], and some methods based on experimental correlations [10].

Yet, great strides have been taken by the recent introduction of what is called as the "engineering transition modeling" by Menter et al. [11] that rely on local data to circumvent some complicated procedures in other methods. Further work demonstrated the viability and practicality of the transition correlation-based model that warranted further investigation. Since the present model [11] does not attempt to model the physics of the transition process, but rather to form a framework for implementation of transition correlations into general purpose CFD methods, different correlations for each user either remain proprietary [11] or user-dependent [12,13] based on the specific experimental data set. Therefore, those correlations that are used in the model are not universal, but reflect each user's own data base. A number of applications for the Menter et al [11] model were done by Genç [14] for a thin airfoil at high Reynolds numbers and Genç, Kaynak and Yapici [15] for flow control around an airfoil by jet blowing and suction at low Reynolds numbers. Following the Menter et al.[11] model, a number of successful two- or three-equation models also appeared in the literature such as k-k_L-ω model of Walters and Leylek [16], near/free-stream intermittency model of Lodefier et al.[17], and k- ω- γ model of Fu and Wang [18]. High speed applications of the Menter et al.[11] model was done by Kaynak [19] for flat plates up to supersonic Mach number of 2.7. Other high speed calculations were done by Fu and Wang [18] for supersonic flow past a straight cone and hypersonic flow over a flared cone at zero angle of attack.

2. Numerical method

In this study, 2-D computational results were obtained using the FLUENT software [20] for NACA64A006 and APEX 16 [4] airfoils. Although 3D effects are present, the aim of this study is to investigate the prediction of the boundary layer and stall characteristics of the airfoils. The k-ω SST turbulence model, k-ω SST transition model and k-k_L-ω transition model are used in conjunction with the built-in RANS solver. The results were compared with the results gathered from the MSES code explained in the study of Drela et al.[4]. The MSES code is a viscous-inviscid interaction code that adapts the Euler equations coupled with the boundary layer code including the e^N method for transition prediction. For both airfoils, free stream boundary conditions were used in the upstream, downstream and outer boundaries. Pressure-farfield boundary condition was used for both cases, with different Mach and Reynolds numbers.

2.1 Solution grid

C-Type structured solution grids were generated by the ICEM CFD software [20] for both NACA64A006 and APEX 16 airfoils as shown in Figures 2 and 3 respectively. The grid extends from -12 chords from upstream to 16 chords downstream and the upper and lower boundary extends 12 chords from the airfoil. Both grids include 61588 cells and 62058 nodes.

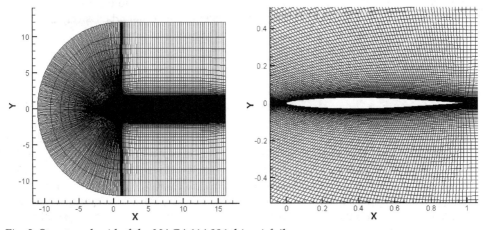

Fig. 2. Structured grid of the NACA64A006 thin-airfoil.

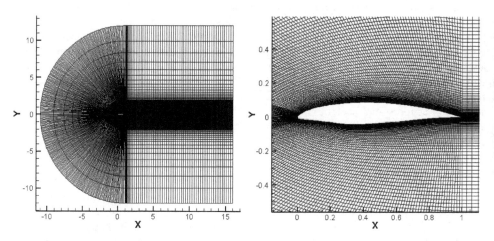

Fig. 3. Structured grid of the APEX 16 airfoil.

2.2 Turbulence and transition model

The k–ω SST turbulence model as implemented within the RANS equations blends the formulation of the Wilcox k–ω model [9] in the near-wall region with the formulation of the k–ε model [21] in the far-field developed by Menter [22]. The k–ω SST transition model [11] solves for four transport equations, such as the turbulent kinetic energy (k), specific turbulence dissipation rate (ω), intermittency (γ), and the transition onset momentum thickness Reynolds number ($Re_{\theta t}$) equations in addition to the basic RANS equations. The γ transport equation and $Re_{\theta t}$ transport equation are used to initiate the transition process and for establishing link with experimental correlations, respectively. The correlations are based on the free stream turbulence intensity (Tu) the $Re_{\theta t}$ at transition onset. The k–k_L–ω model [16] solves three transport equations for the turbulent kinetic energy (k), laminar kinetic energy (k_L), and specific turbulence dissipation rate (ω) in addition to the RANS equations. The k_L is based on the non-turbulent fluctuations in the laminar boundary layer, as defined in the work of Mayle and Schulz [23].

2.3 Flow cases

For the NACA64A006 airfoil, the numerical results are compared with the results of experiments conducted by McCollough and Gault [24]. In those experiments, Mach number and Reynolds number were 0.17 and 5.8 million respectively, and the angle of attack ranged from 2°to 10°. For the APEX 16 airfoil, the numerical results obtained using the FLUENT [20] k–k_L–ω transition model are compared with the numerical results obtained using the MSES code by Drela [4]. In the first part of the simulations, Reynolds numbers of 200,000, 300,000 and 500,000 are set at a constant Mach number of 0.6. In the second part of the simulations, Reynolds numbers of 200,000, 300,000 and 500,000 are simulated for Mach numbers 0.6, 0.65 and 0.70 for each Reynolds number. For all cases, the angle of attack ranges from -4° to 8°.

3. Results and discussion

3.1 NACA64A006

Figure 4 shows the numerical data obtained by using the FLUENT [20] showing experimental C_L and C_D values for different angle of attack values for the NACA64A006 airfoil. In this figure, all models give reasonably good results against the experiment in the linear region, but the calculation starts to differ from the experiment [24] after 5º angle of attack. It is observed that after 5º, the k-ω SST and k-ω-SST transition models underpredict the lift coefficient, whereas the k-k_L-ω transition model predicts the lift coefficient better than those two models. On the other hand, Figure 4 shows that there is a quite good agreement between the experiment and computational results for the drag coefficient.

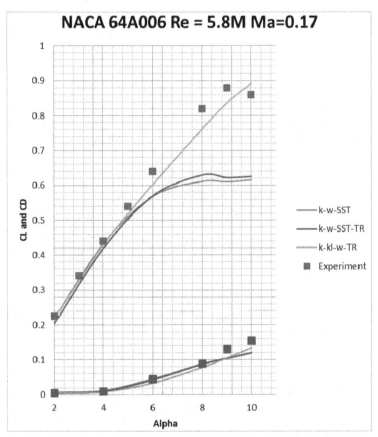

Fig. 4. Lift and drag coefficients of the NACA64A006 airfoil at different angles of attack.

Figure 5 shows the pressure coefficient distributions at 4 degrees angle of attack where the experimental and numerical data are nearly the same. At this angle of attack, pressure coefficient distribution does not indicate any flow separation. However at 6 degrees of angle of attack, there is a hump near the leading edge of the airfoil, which indicates a separation bubble with a size of about 20% of the chord length. At 9 degrees of angle of attack, Figure 5 shows that there are significant differences between the experiment and numerical predictions in the pressure coefficient distribution. Comparing Figs. 4 and 5, although the k-k$_L$-ω transition model appears to yield the best result for the lift coefficient, detailed pressure distributions do not wholly support this finding as there are large discrepancies for the local pressures.

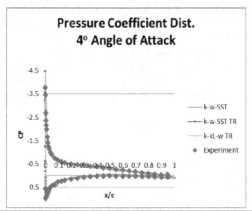

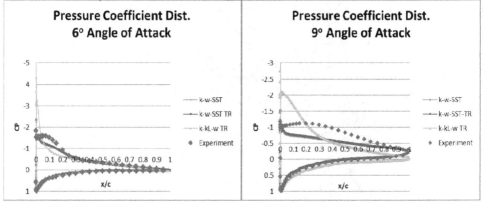

Fig. 5. Numerical and experimental pressure coefficient distributions of the NACA64A006 airfoil at different angles of attack.

3.2 APEX 16

Numerical results for APEX 16 airfoil are divided into two categories such as one high subsonic Mach number at a range of low Reynolds numbers and a range of low Reynolds numbers at range of high subsonic Mach numbers. All the numerical data for this airfoil is gathered using k-k$_L$-ω transition model and they are compared with the numerical results obtained using the MSES code mentioned in the work of Drela et al [4].

i.   High Subsonic Mach Number Case:

In this section, a constant Mach number of 0.6 is selected and the simulations are made at Reynolds numbers of 200,000, 300,000 and 500,000. The aim of this investigation is to understand the effect of Reynolds number for the aerodynamic characteristics of the APEX

16 airfoil. Also in this section, prediction performances of the MSES code and k-k$_L$-ω transition model are compared.

Figure 6 shows the numerical data of C$_L$ values against C$_D$ values using the k-k$_L$-ω transition model and MSES code for different Reynolds numbers. This figure shows that the MSES code and k-k$_L$-ω transition model reasonably agree for the lift coefficient; but, there are some differences in the drag coefficient. Especially for low Reynolds numbers, the k-k$_L$-ω transition model predicts smaller drag coefficients than the MSES code. As the Reynolds number increases, the agreement between two models start improving.

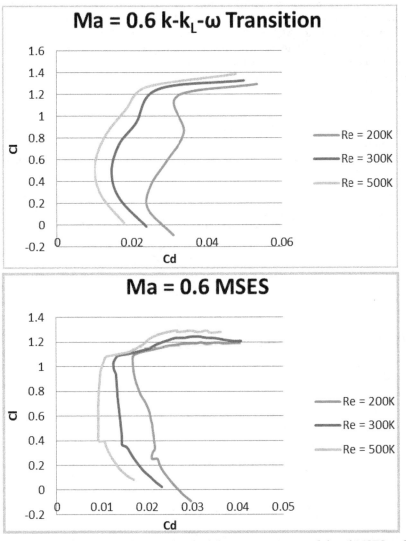

Fig. 6. Comparison of C$_L$-C$_D$ predictions for the k-k$_L$-ω transition model and MSES code for the APEX 16 airfoil at Ma = 0.6.

For low Reynolds number flow conditions, predicting the drag coefficient becomes even more difficult, since low Reynolds number airfoils typically exhibit laminar separation bubbles, which are known to significantly affect the performance of an airfoil. As for increasing the Reynolds number, the ability of prediction of the drag and lift coefficient becomes easier for the flow solvers because the flow is no longer laminar, and turbulent boundary layer is effective on the surface of the airfoil beginning from the leading edge.

From Figure 6, it can be conluded that as the Reynolds number gets higher, lift coefficient gets higher whereas the drag coefficient gets lower. Although the predictions of MSES code and k-k_L-ω transition model do not fully agree, the relative agreement is still reasonable as both models at least agree on the trend of the lift and drag coefficients as the Reynolds number changes. Taking Ma = 0.6 and Re = 300,000, further information on the characteristics of the APEX 16 airfoil is obtained using FLUENT's k-k_L-ω transition model and the results are compared with the MSES code. For this condition, pitching moment predictions, lift coefficient predictions and transition location on the upper surface of the airfoil are compared.

Figure 7 shows the lift coefficient predictions versus angle of attack comparison for the Ma = 0.6 and Re = 300,000 case. There is a good agreement for the lift coefficient between the k-k_L-ω transition model and MSES code predictions until the angle of attack reaches around 5 degrees; but after 5 degrees, the MSES code predicts lift coefficient lower than k-k_L-ω transition model.

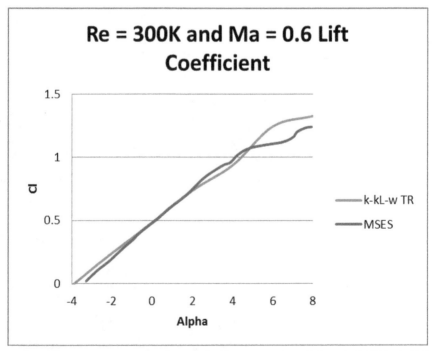

Fig. 7. Lift coefficient versus angle of attack for Re = 300,000 and Ma = 0.6 MSES and k-k_L-ω transition model comparison.

Figure 8 shows the pitching moment predictions for the same case. The pitching moment results obtained using FLUENT are all based on the pitching moment center at 25% of the chord length. The results show that, there is a good agreement between the two models in the angle of attack range from 0 to 4 degrees. However, the k-k$_L$-ω transition model calculates a larger pitching moment for angles of attack larger than 4 degrees than the MSES code.

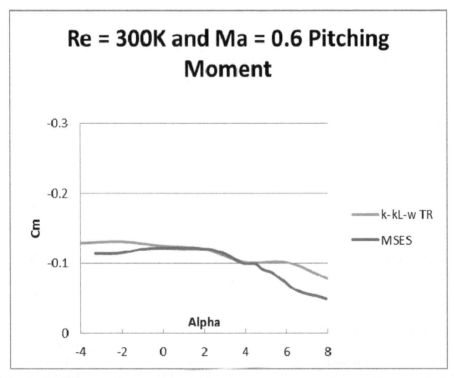

Fig. 8. Pitching moment versus angle of attack for Re = 300,000 and Ma = 0.6 MSES and k-k$_L$-ω transition model comparison.

Finally, Figure 9 shows the predictions of the location of the transition location on the upper surface of the airfoil against the lift coefficient. The transition locations obtained from the FLUENT results are all assumed that the transition occurs when turbulent to laminar viscosity ratio reaches about 80-100. Looking at this figure, the agreement between the two models are very good at low angles of attack. The reason why both curves are not matching perfectly is that, the lift coefficient predictions are harder to obtain at high angles of attack.

ii.   High Subsonic Mach Numbers at a Range of Low Reynolds Numbers

In this section, low Reynolds numbers in the range (200,000-500,000) are kept constant, and Mach numbers are changed in the high subsonic range of 0.60, 0.65 and 0.70 at each Re number in order to understand the effect of changing Mach number.

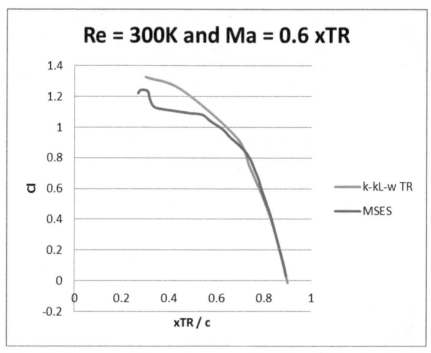

Fig. 9. Transition location prediction on the upper surface of the airfoil against the lift coefficient comparison at Re = 300,000 and Ma = 0.6.

3.3 Re = 200,000 case

Figure 10 shows the high subsonic Mach number data for the C_L against C_D values gathered from k-k_L-ω transition model and the MSES code at Re=200,000. As seen in this figure, the drag coefficients calculated by the k-k_L-ω transition model are much higher than the MSES code. Also, for Mach number 0.7, lift coefficient calculations of k-k_L-ω transition model and the MSES code differs a lot. However, the k-k_L-ω transition model which is based on Reynolds-averaged Navier-Stokes equations seems to produce smoother drag polars which more look like normal trend. On the contrary, the MSES code predicts smaller C_D values as C_L gets larger which should not be a normal pattern.

3.4 Re = 300,000 case

Figure 11 shows the high subsonic Mach number data for the C_L against C_D values gathered from k-k_L-ω transition model and MSES code at Re=300,000. As seen in the figure, the drag coefficients calculated by k-k_L-ω transition model were again much higher than the MSES code. Also, for Mach number 0.7, lift coefficient calculations of k-k_L-ω transition model and the MSES code differs a lot. Yet, as Reynolds number gets larger in this case, the shape of the drag polars slightly start normalizing as they look more like standard high Re number drag polars. Some kind of kink is apparent in the MSES simulations around C_L=0.4.

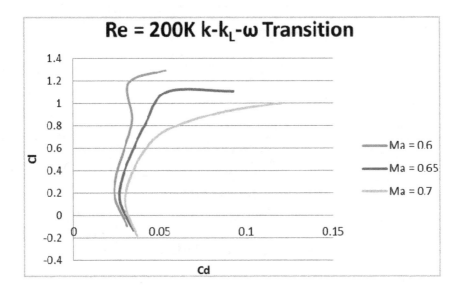

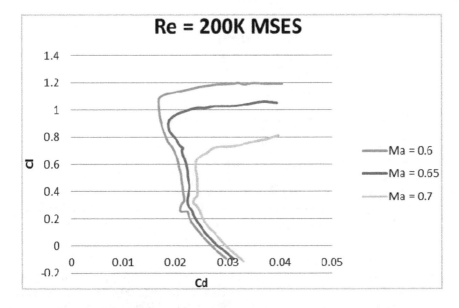

Fig. 10. Comparison of C_L-C_D predictions of k-k_L-ω transition model and MSES code for APEX 16 airfoil at Re = 200,000.

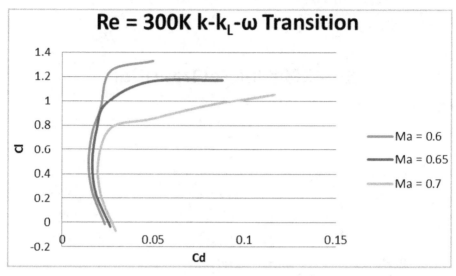

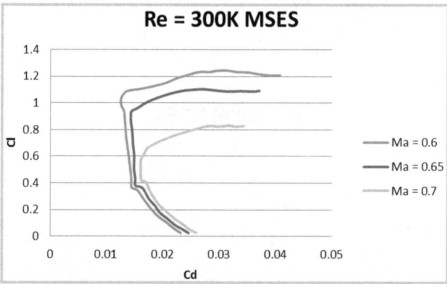

Fig. 11. Comparison of C_L -C_D predictions of k-k_L-ω transition model and MSES code for APEX 16 airfoil at Re = 300,000.

3.5 Re = 500,000 case

Figure 12 shows the high subsonic Mach number data for the C_L against C_D values gathered from k-k_L-ω transition model and the MSES code at Re=500,000. The drag coefficients calculated by the k-k_L-ω transition model were again much higher than the MSES code. The shape of the drag polars quickly start normalizing as they look more like standard high Re number drag polars. The kink which is apperant in the MSES simulations around C_L=0.4 is

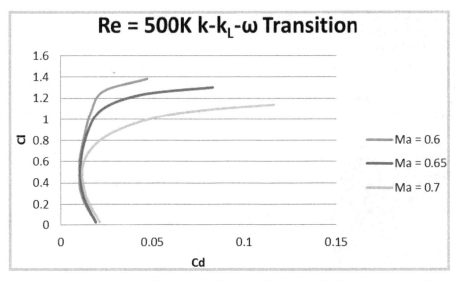

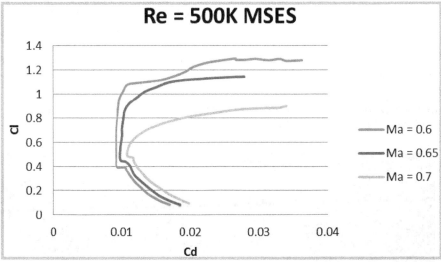

Fig. 12. Comparison of C_L -C_D predictions of k-k_L-ω transition model and the MSES code for the APEX 16 airfoil at Re = 500,000.

more pronounced for this Re number. From all these numerical analysis for the APEX 16 airfoil, it can be concluded that the drag coefficient decreases as Reynolds number increases for constant Mach numbers, and also drag coefficient increases as Mach number increases for the same Reynolds number. It appears that as the Mach number increases, the adverse pressure gradient causes the drag coefficient to increase while keeping the Reynolds number constant.

The separation bubble stability has an important effect on the airfoil predicted performance [4]. The MSES code, based on the stable bubble calculations, predicts a lift coefficient of 0.96 at the flight condition of Ma = 0.65, Re = 200,000 and an angle of attack of 4 degrees, whereas k-k$_L$-ω transition model predicts an average section lift coefficient of 0.82 for the same flight condition. In the following, the shape, location and extent of the separation bubbles for different flow cases are introduced.

Figure 13 shows the velocity contours and the velocity vectors of the APEX 16 airfoil at 0° angle of attack, Ma = 0.6 and Re = 300,000 based on k-k$_L$-ω transition model. From this figure, it is seen that the flow separation occurs at around 0.60c away from the leading edge. At this angle of attack, flow reattachment is not observed until the trailing edge. Figure 14

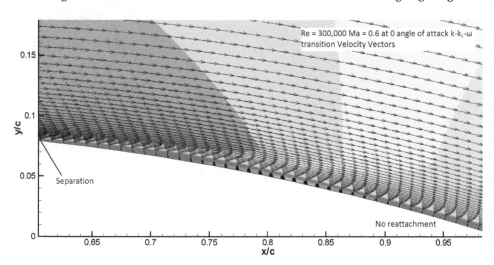

Fig. 13. Velocity contours and velocity vectors of the APEX 16 airfoil at 0° angle of attack, Ma = 0.6 and Re = 300,000 based on k-k$_L$-ω transition model.

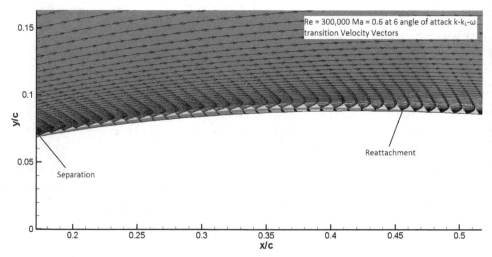

Fig. 14. Velocity contours and velocity vectors of the APEX 16 airfoil at 6° angle of attack, Ma = 0.6 and Re = 300,000 based on k-k$_L$-ω transition model.

shows the velocity contours and the velocity vectors of the APEX 16 airfoil at 6° angle of attack, Ma = 0.6 and Re = 300,000 based on k-k$_L$-ω transition model. From this figure, it is seen that the flow separation occurs at around 0.10c away from the leading edge. Comparing this figure with Figure 13, it is concluded that the flow separation occurs closer to the leading edge as the angle of attack increases, as expected.

Figure 15 shows the turbulent kinetic energy distributions for 0° and 6° angle of attack for the APEX 16 airfoil at the same flow conditions. This figure supports the observations and comments made for Figures 13 and 14. As seen in Figure 15, the flow transition into turbulence occurs closer to the leading edge as the angle of attack increases. Also, there is a laminar boundary layer separation first, followed by a shear layer transition into turbulence, and finally there is turbulent reattachment.

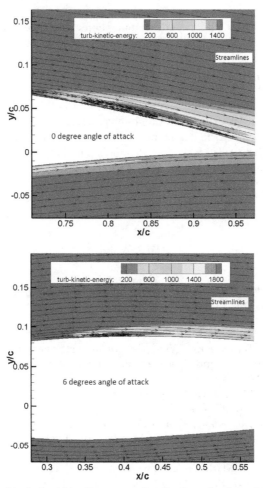

Fig. 15. Comparison of turbulent kinetic energy distributions for 0° and 6° angle of attack for the APEX 16 airfoil.

4. Conclusions

In this study, firstly, aerodynamic characteristics of the NACA64A006 airfoil is investigated using the k-ω SST turbulence model, k-ω SST transition model and k-k$_L$-ω transition model using FLUENT. The results obtained from FLUENT are compared with the experiment [24], and it is observed that all numerical approaches give reasonably good results in the linear region, although the results began to differ as the angle of attack gets larger. Especially after 5° of angle of attack, whereas the k-ω SST turbulence and k-ω SST transition models greatly underpredict the lift coefficient, the k-k$_L$-ω transition model yields the best result. With regards to the drag coefficient, it is reasonable to say that all numerical methods agree quite well with the experimental data. For the pressure coefficient, it is observed that the k-k$_L$-ω transition model also fares better than the other models.

In the second part of this study, aerodynamic characteristics of the APEX 16 airfoil is investigated using the k-k$_L$-ω transition model and results are compared with the e^N based results of the MSES code by Drela et al [4]. The first apparent characteristic for the APEX 16 is that increasing the Mach number results in a decrease in the maximum lift coefficient. Although the lift coefficient predictions of the MSES code and k-k$_L$-ω transition model slightly differ, the general trends in both results are similar. As the Reynolds number decreases, the separation bubbles become larger, which is the reason for the increase in the drag coefficient. Another observation is that the slope of the lift curve is relatively unaffected by the Mach and Reynolds numbers except near the stall. The predicted transition location versus the lift coefficient is also presented for the Re = 300,000 and Ma = 0.6 case where both models agree on the transition location but the lift coefficient predictions differ at high angles of attack. It is shown that the transition location on the upper surface moves forward with the increasing angle of attack, as expected.

5. References

[1] "Fixed and Flapping Wing Aerodynamics for Micro Air Vehicle Applications", ed. T.J. Mueller, Vol. 195, Progress in Astronautics and Aeronautics, AIAA, Inc. VA, USA.

[2] E. L. Fleming, S. Chandra, M. R. Shoeberl, and J. J. Barnett, "Monthly Mean Global Climatology of Temperature, Wind, Geopotential Height and Pressure for 0-120 km", National Aeronautics and Space Administration, Technical Memorandum 100697, Washington, D.C., 1988.

[3] Marchmann, J. F., "Aerodynamic testing at low Reynolds numbers", *J. Aircraft*, vol. 24, no. 2, Feb. 1987, pp. 107-114

[4] Greer, D., Hamory, P., Krake, K. and Drela, M., "Design and Predictions for a High-Altitude (Low-Reynolds-Number) Aerodynamic Flight Experiment", NASA/TM-1999-206579.

[5] Giles, Michael B. and Mark Drela, "Two-Dimensional Transonic Aerodynamic Design Method,"*AIAA Journal*,vol. 25, no. 9, Sept. 1987, pp. 1199–1206.

[6] Drela, Mark and Michael B. Giles, "Viscous-Inviscid Analysis of Transonic and Low Reynolds Number Airfoils," *AIAA Journal*, vol. 25, no. 10, 1987, pp. 1347–1355.

[7] Tatineni, M. and Zhong, X., "Numerical Simulation of Unsteady Low-Reynolds-Number Separated Flows Over Airfoils," AIAA 97-1929, July 1997.

[8] Tatineni, M. and Zhong, X, "Numerical Simulations of Unsteady Low-Reynolds-Number Flows Over the APEX Airfoil," AIAA 98-0412, Jan. 1998.

[9] Wilcox, D.C., "Simulation of Transition with a Two-Equation Turbulence Model", *AIAA Journal*, Vol. 32, No. 2, 1994, pp. 247-255.

[10] Suzen, Y. B. and Huang, P.G., "Modeling of Flow Transition Using an Intermittency Transport Equation", ASME *Journal of Fluid Engineering*, Vol. 122, pp. 273-284 (2000).

[11] Menter, F.R., Langtry, R.B., Likki, S.R., Suzen, Y.B., Huang, P.G. and Völker, s. "A Correlation Based Transition Model Using Local Variables: Part I-Model Formulation ASME-GT2004-53452, Proceedings of ASME Turbo Expo 2004, Vienna, Austria, 2004, pp. 57-67.

[12] Misaka, T. and Obayashi, S. "A Correlation-Based Transition Models to Flows Around Wings," AIAA Paper 2006-918.

[13] Gürdamar, E., Çete, R, Aksel, H. and Kaynak, Ü, "Improved Transonic Wing Flow Calculations Using Transition Correlations", AIAA Paper 2006-3170.

[14] Genç, M.S., "Numerical simulation of flow over a thin aerofoil at a high Reynolds number using a transition model", Proc. IMechE, Part C - J. Mechanical Engineering Science, Vol. 224 (10), pp. 2155 - 2164, 2010.

[15] Genç, M.S., Kaynak, Ü. and Yapıcı, H.,"Performance of transition model for predicting low Re airfoil flows without/with single and simultaneous blowing and suction", European Journal of Mechanics B/Fluids, Volume 30, Issue 2, March-April 2011, Pages 218-235

[16] Walters, D.K. and Leylek, J.H., "A New Model for Boundary-Layer Transition Using a Single Point RANS Approach," *ASME J. Turbomach.* 2004; 126: 193-202.

[17] Lodefier, K., Merci, B., De Langhe, C., and Dick, E. "Transition Modeling with the SST Turbulence Model and Intermittency Transport Equation," ASME Turbo Expo, 2003, Atlanta, GA, USA.

[18] Fu, S. and Wang, L. "A Transport Intermittency Model for Supersonic/Hypersonic Boundary Layer Transition,"5th. European Congress on Computational Methods in Applied Sciences and Engineering (ECCOMAS 2008), June 30 – July 5, 2008, Venice, Italy.

[19] Kaynak, Ü., "Supersonic Boundary-Layer Transition Prediction under the Effect of Compressibility Using a Correlation Based Model", Proc IMechE, Part G-Journal of Aerospace Engineering, *accepted on 16 June 2011*, DOI: 10.1177/0954410011416187.

[20] Fluent Incorporated. Fluent (V 6.4) User's Guides, 2007.

[21] Launder, B. E. and Spalding, D. B., "Lectures in mathematical models of turbulence", 1972, Academic Press, London, England

[22] Menter, F., "Two-equation Eddy Viscosity Turbulence Models for Engineering Applications", *AIAA Journal*, Vol. 32, No. 8, 1994, pp. 1598-1605.

[23] Mayle, R. E. and Schulz, A. "The path to predicting bypass transition." *ASME J. Turbomach.*, 1997; 119: 405-411.

[24] McCollough, G. B. and Gault, D. E. "Boundary-layer and stalling characteristics of the NACA64A006 airfoil section." NACA TN1923, 1949.

4

Prediction of Aerodynamic Characteristics for Elliptic Airfoils in Unmanned Aerial Vehicle Applications

Varun Chitta, Tej P. Dhakal and D. Keith Walters

Mississippi State University,
Starkville, MS,
USA

1. Introduction

Since the mid 1920's, when the first attempt was made to fly a powered pilotless vehicle (Newcome, 2004), significant design improvements for unmanned aerial vehicles (UAV) have been developed, motivated by increased dependence on these vehicles by both civilian and military organizations. Today, the widespread use of UAV's and recent advances in technology resulted in greater interest than ever before for research on these vehicles.

An unmanned vehicle with an additional capability of vertical take-off and landing (VTOL) represents one example of an interest area which has potential for significant research innovation. A canard rotor/wing (CRW) is a UAV that falls under this category. CRWs can hover and fly at low speeds like a conventional helicopter and can also fly at high speeds like a fixed wing aircraft with the additional capability of VTOL (Rutherford et al., 1993; Pandya & Aftosmis, 2001). The CRW has a stoppable two bladed rotor design which allows it to take-off vertically from the ground, transition to a fixed wing aircraft by locking its rotor, and cruise at higher speeds. This specific ability of CRWs to transform into various flight modes makes them an interesting option for military and civilian applications. However, the transition from a rotor blade to a fixed wing vehicle takes place at low speeds, and requires the cross section of rotor blades be elliptic (Kwon & Park, 2005). These considerations motivate research into the aerodynamic characteristics of elliptic airfoils at low/transitional Reynolds numbers (Re).

For lifting surfaces of conventional aircraft, Re is typically well above 10^6 and the turbulent boundary layer does not separate until high angles of attack are encountered (Jahanmiri, 2011). In contrast, UAVs have lower flight velocities and are smaller in size, which results in low wing chord Reynolds numbers ($10^5 < Re < 2 \times 10^6$) that often lie in the transitional regime. It is well known that, for low Re flows, viscous effects play a much more important role than in high Re flows, in which viscous effects are either neglected or restricted to a thin region near body surface. The complex interactions of viscous mechanisms, transition, and separation present an interesting and challenging problem for UAV design.

For low freestream turbulence intensity (FSTI) and low Re flows, boundary layers are initially laminar and are prone to separation under the influence of even mild adverse pressure gradients. Once separated, the laminar boundary layer forms a shear layer that may quickly undergo transition to turbulence and reattach to the airfoil surface in the form of a turbulent boundary layer, leading to the formation of a laminar separation bubble (LSB) (Jones, 1938; Diwan & Ramesh, 2007). Shear layer transition occurs due to the amplification of flow instabilities, which cause the shear layer to roll up and form vortices that play a vital role in bubble formation. The enhanced momentum transport in turbulent flow enables flow reattachment and results in development of a turbulent boundary layer on the downstream portion of the airfoil (Sandham, 2008).

The post separation behavior of laminar boundary layers is quite interesting, and accounts for a deterioration of aerodynamic performance of low Re airfoils which is exhibited by an increase in drag and decrease in lift (Yang & Hu, 2008). Experimental results show that airfoil performance starts to deteriorate when chord Re decreases below 5×10^5 (Lissaman, 1983; Carmichael, 1981). Also, if Reynolds numbers are below 5×10^4, the separated shear layer may fail to reattach to the airfoil surface, resulting in a large wake region behind the body (Lin & Pauley, 1996; Yarusevych & Sullivan, 2006) and dramatic loss of performance.

Gaster (Gaster, 1967) was the first to study transition of laminar separation bubbles. As described by Gaster, the separated shear layer formed after laminar boundary layer separation from the suction surface of an airfoil may reattach back to the surface, thereby forming a shallow region of reverse flow known as the separation bubble. A "dead air" region of low velocity is observed under the detached shear layer immediately after separation which results in the formation of a nearly constant region of pressure on the airfoil surface. A strong recirculation zone is observed near the downstream region of bubble. Momentum transfer due to turbulent mixing eliminates the reverse flow due to entrainment of high speed outer fluid, and finally flow reattaches to suction surface (Jahanmiri, 2011).

Depending on the size of bubble, LSB's are typically categorized as either short or long bubbles (Tani, 1964). A long bubble occupies a significant portion of the airfoil surface and affects the inviscid pressure and velocity distributions over much of the airfoil, whereas a short bubble covers only a small portion of airfoil surface and does not affect the pressure and velocity distributions. Figure 1 shows the velocity vector plot of airfoil for SST k-ω model (Menter, 1994) at angle of attack (α) = $7°$. For this case, no separation bubble is formed over suction surface of airfoil as the flow is turbulent throughout the surface of airfoil and separation occurs only near the trailing edge. Figure 2 shows a LSB formed on the suction surface near the leading edge of airfoil at $\alpha = 7°$ using the transition-sensitive k-k_L-ω (Walters and Cokljat, 2008) model. The presence of a long separation bubble and post separation behavior of the boundary layer results in increased drag and decreased lift coefficient (Lian & Shyy, 2007).

At high angles of attack, the separated shear layer may fail to reattach to the suction surface and either a long bubble or a completely unattached free shear layer may develop (Gaster, 1967). The change in flow reattachment process due to increasing α depends on the type of airfoil and flow conditions, and might occur gradually or quite sharply. Bubble bursting occurs as a fundamental breakdown of the flow re-attachment process (Horton, 1967). The bursting of the bubble creates an increase in drag, undesirable change in pitching moment and an appreciable drop in lift, causing the airfoil to stall. Figures 3, 4 and 5 show velocity contours of an elliptic airfoil obtained using the k-k_L-ω model at $\alpha = 0°$, $7°$ and $18°$ respectively. At $\alpha = 0°$, flow separation is observed on the downstream of portion of the airfoil followed by the formation of two symmetric vortices near the trailing edge. At this angle, the flow is symmetrical over the suction and pressure sides. At $\alpha = 7°$, a laminar separation bubble is observed on the suction surface near the leading edge of the airfoil. The flow transitions and reattaches as a turbulent boundary layer, and a second flow separation is observed at a location farther downstream on the airfoil surface as compared to the flow separation for the $\alpha = 0°$ case. At $\alpha = 18°$, bubble bursting takes place and results in a reverse flow over the entire suction surface of airfoil, indicating that the airfoil has stalled.

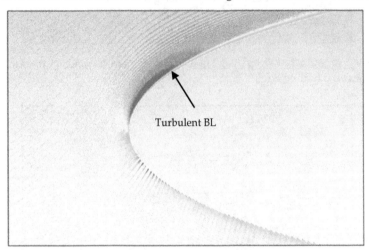

Fig. 1. Velocity vectors near the leading edge of an elliptic airfoil for SST k-ω model at $\alpha = 7°$. For this case, no separation bubble is observed on the suction surface.

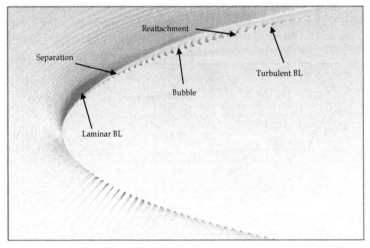

Fig. 2. Laminar separation bubble visible on the suction surface near the leading edge of an elliptic airfoil. Results obtained for k-k_L-ω model at $\alpha = 7°$.

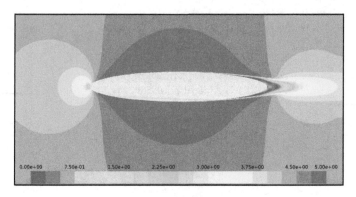

Fig. 3. Velocity contours for k-k_L-ω model at $\alpha = 0^0$. Two opposite and symmetrical vortices are formed near the trailing edge.

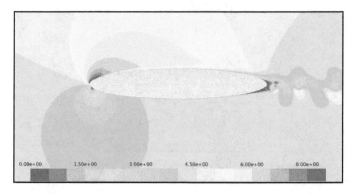

Fig. 4. Velocity contours for k-k_L-ω model at $\alpha = 7^0$. A laminar separation bubble is formed on the suction surface near the leading edge of the airfoil. Vortex shedding is captured by k-k_L-ω for this case and is observed near the trailing edge and in the wake region of airfoil.

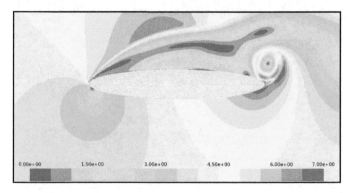

Fig. 5. Velocity contours for k-k_L-ω model at $\alpha = 17^0$. Bubble bursting has occurred and reverse flow over the entire suction surface is observed.

This chapter presents CFD simulations of static elliptic airfoils at varying angles of attack and at relatively low Reynolds numbers ($10^5 <$ Re $< 2 \times 10^6$), and compares the numerical results with available experimental data (Kwon & Park, 2005). In an effort to facilitate improved understanding of transitional and turbulent flow physics, we have performed numerical simulations using several commercially available fully turbulent and transition-sensitive RANS models, as well as a curvature sensitive fully turbulent RANS model recently developed by our group. Results indicate that a transition-sensitive model is required to accurately reproduce the separation bubble that appears on the suction surface near the leading edge of airfoil over a relatively large range of angles of attack prior to stall. Although the transition-sensitive models — k-k_L-ω (Walters and Cokljat, 2008) and transition-sensitive SST (Menter et al., 2004) — produced reasonable results, both models failed to accurately predict the stall point of airfoil. In contrast, the curvature sensitive SST k-ω-v^2 fully turbulent model (Dhakal and Walters, 2011) predicted the stall point close to experimental results, but it failed to accurately predict the transitional characteristics including flow separation and reattachment in the leading edge region. A comprehensive turbulence model considering both curvature and flow transition effects of airfoil at low Re is not yet available, but our results suggest that such a model would be highly desirable for solving fluid flow problems faced by elliptic airfoils.

The chapter is organized as follows. In the next section we discuss several recent experimental and computational studies related to elliptical bodies in order to provide context for our effort; In section 3 we introduce the computational methodology employed in this work and give the geometric description of the chosen airfoil and the relevant flow parameters. Section 4 presents the numerical results from CFD simulations and provides a discussion on the outcome. In the results section, we also validate the CFD results against available experimental data, and highlight the reasons for discrepancies between experimental and computational results. Section 5 provides a summary and conclusions.

2. Literature review

There are relatively few experimental and computational studies available in the open literature regarding the study of aerodynamic characteristics of elliptic airfoils at transitional Reynolds numbers. This section presents a brief summary of publications most relevant to the current study.

2.1 Experimental studies

Zahm et al. (1929) reported wind tunnel test results for four elliptic cylinders with fineness ratios of 2.5, 3.0, 3.5 and 4.0. Surface pressures and drag characteristics were studied for various yaw angles. Zahm found that, for low Re flows, optimal drag characteristics occurred when the elliptic cylinder had a fineness ratio of 4.0, whereas for high Re flows, improved characteristics were obtained for fineness ratios smaller than 4.0. Schubauer (1939) studied the air flow in the boundary layer on an elliptic cylinder. A conventional hot-wire anemometer was used to measure magnitude and frequency of speed fluctuations in the boundary layer. The study investigated the relationship between boundary layer transition and freestream turbulence intensity for a 33.8% thickness elliptic cylinder at zero angle of attack. Schubauer found that the transition location depended on both the turbulence scale and the freestream turbulence intensity.

Kwon and Park (2005) performed wind tunnel tests for an elliptic airfoil and measured aerodynamic forces and moments for a single airfoil thickness ratio of 16% at Re = 3 × 10⁵. Tests were conducted on an airfoil with and without a boundary layer transition trip with FSTI of 0.12%. Trip devices were attached on both the pressure and suction surfaces of the airfoil at about 10% chord length to induce turbulent flow over the majority of the surface. The boundary layer trip technique is generally used in wind tunnel tests to simulate full scale or high Re flows in low Re airfoil test conditions in a laboratory to enforce transition locations and to eliminate laminar separation bubbles (Kwon et al., 2006). It was found that the lift curve of the elliptic airfoil varied as function of Re and lift did not linearly increase with angle of attack, in contrast to the behavior of conventional airfoils. In the experiment, C_L curves for both smooth and tripped cases behaved similarly when the angle of attack exceeded 6°. Kwon also found that the asymmetric flow separation behavior around the smooth airfoil trailing edge caused a lift curve slope much greater than 2π at low angles of attack which differs from the behavior of conventional airfoils.

Kwon et al. (2006) extended the previous research on elliptic airfoils to study in detail the boundary layer transition process using a particle image velocimetry (PIV) technique. Tests were conducted on the same elliptic airfoil as in Kwon and Park (2005) for the same flow conditions. Velocity profiles were measured and shape factors were calculated from PIV measurements. Intermittency factors were computed from surface mounted hot film sensor measurements. The authors concluded that the unusual aerodynamic characteristics of elliptic airfoils, such as a high lift curve slope and high drag coefficient at low angle of attack, were a consequence of the different flow regimes, i.e. laminar or turbulent, between the suction and pressure surfaces as angle of attack increases.

2.2 Computational studies

Johnson et al. (2001) used a 2D spectral element method to solve the unsteady Navier-Stokes equations to study the vortex structures behind two-dimensional elliptic cylinders. The effects of Re and aspect ratio on Strouhal number, drag coefficient and the onset of vortex shedding were reported for Reynolds numbers from 30 to 200 and aspect ratios ranging from 0.01 to 1. As the aspect ratio of the elliptic cylinder was decreased, the shedding pattern behind the cylinder changed from a periodic Karman vortex street to an aperiodic secondary shedding of vortices. The value of Re at the onset of periodic vortex shedding decreased as aspect ratio decreased. In general, however, this range of Re is too low to be directly applicable in UAV design.

Kim and Sengupta (2005) focused their computational study on the unsteady viscous flow over two dimensional elliptic cylinders by solving the incompressible Navier-Stokes equations for thickness-to-chord ratios of 0.6, 0.8, 1.0 and 1.2, and Re ranging from 200-1000. The total drag force on elliptic cylinders during unsteady viscous airflow mostly comes from the pressure drag force, which increases with an increase of either thickness-to-chord ratio or Re. Also, the mean pressure drag force strongly depends on cylinder thickness, while the mean frictional drag force strongly depends on Re. The frequency of vortex shedding was found to be higher when either the thickness of the elliptic cylinder was reduced or when Re was increased. The authors concluded that both thickness-to-chord ratio and Re have significant effects on vortex shedding and also on the amplitudes of lift and drag forces.

Assel (2007) performed a computational study of flow over elliptic airfoils for a range of Re from 1×10^5 to 8×10^6, by varying the thickness ratios of the airfoils from 5% to 25% and angles of attack from 0 to 20°. He used the Spalart-Allmaras (SA) turbulence model to perform steady-state CFD simulations for the test case. For a Re of 3×10^5 and a thickness ratio of 16%, Assel compared his CFD results with experimentally available wind tunnel test results (Kwon & Park, 2005). Although the stall point of airfoil was reported to be predicted accurately, transitional effects on the airfoil in the CFD simulations did not match with experimental results. The formation of a laminar separation bubble was observed at $\alpha = 8°$ in the CFD simulations, while Kwon and Park reported the occurrence of transition over the airfoil surface for the smooth case at 6°. Also, CFD results did not possess the unusual aerodynamic characteristics of elliptic airfoils such as high lift curve slope and high drag coefficient at low angle of attack ranges. These discrepancies are likely due in whole or in part to the inability of fully turbulent models to resolve transition effects correctly.

Pandya & Aftosmis (2001) studied the aerodynamic loads on a CRW aircraft using inviscid numerical simulations to understand flight characteristics during conversion from rotary to fixed-wing flight. Although the authors studied the loads acting on different components of the aircraft, little attention was given to the detailed analysis of the fluid mechanics and aerodynamic forces acting on lifting surfaces of aircraft.

To date, CFD simulations of flow over elliptic airfoils have been performed using traditional eddy viscosity models that were developed primarily for high Re applications. Such models are mostly used for predicting fully turbulent flow in which transition effects and rotation and/or curvature effects do not significantly affect the mean flow. Complex flow phenomena like formation of laminar separation bubbles and flow transition from laminar to turbulent are quite commonly encountered in applications of low Re flows. Usage of standard, fully turbulent models for these applications may lead to accuracy degradation in the prediction of flow characteristics as these models do not have the ability to accurately predict the transitional behavior of fluid flow. Recently, the laminar kinetic energy (LKE) concept has led to the recent development of RANS based turbulence models intended to capture the flow transition effects at low Re flows without the use of intermittency factors (Walters, 2009).

An early version of an LKE based model was introduced by Walters & Leylek (2004), which provided a single point RANS approach for transitional flow prediction which eliminates the need for an external linear stability solver or empirical transition correlations. The most recently documented version of the model is the k-k_L-ω model (Walters & Cokljat, 2008). It is a three-equation eddy viscosity model which has transport equations for turbulent kinetic energy (k), laminar kinetic energy (k_L) and specific dissipation rate (ω). As an alternate approach, the Transition SST k-ω model (Menter et al., 2004) has also been introduced as a single-point approach for transitional flow prediction. It is a four-equation model, with two additional transport equations beyond k and ω; one to determine intermittency (γ – equation) and one to determine the transition onset momentum thickness Re ($Re_{\theta t}$ - equation). Recently, Genc et al. (2009, 2011) performed detailed studies to evaluate the performance of the transition-sensitive k-k_L-ω and Transition SST k-ω models versus fully turbulent k-ω SST (Menter, 1994) and k-ε RNG (Choudhury, 1993) models for predicting low Re flows over a NACA 2415 airfoil for a flow Re of 2×10^5. It was shown that both transition-sensitive models improve predictive capability over fully turbulent model form, although differences between the transition-sensitive models were noted.

Most eddy viscosity models also fail to accurately predict the effects of system rotation or streamline curvature, which can enhance or reduce the turbulence intensity in attached boundary layers and separated shear layers. As discussed in the recent review of curvature-sensitized RANS models (Durbin, 2011), convex curvature tends to reduce turbulence intensity while concave curvature tends to enhance it. These effects of curvature are determined by the direction of rotation: along a convex wall, the strain rate tensor rotates in the same direction as the local vorticity vector; along a concave wall, the two rotations are in opposite directions. Co-rotation suppresses turbulence and counter-rotation enhances it. Recently, a new model sensitized to system rotation and streamline curvature was introduced by Dhakal and Walters (Dhakal & Walters, 2011). The model was dubbed SST k-ω-v^2, and includes terms to modify the eddy viscosity in response to local curvature of the mean flowfield.

In this study, both of the transition-sensitive models discussed above—k-k_L-ω and Transition SST k-ω model—are used. Both of these models are incorporated into Ansys FLUENT and are therefore commercially available. These models are used to evaluate the importance of resolving boundary layer transition for analysis of aerodynamic characteristics of low Re elliptic airfoils. Furthermore, results are compared to experimental data in order to validate the capability of the models for accurate prediction of flow physics. Similarly, to evaluate the impact of flow curvature effects on the aerodynamic characteristics of an elliptic airfoil, simulations have also been carried out in this study using the newly developed SST k-ω-v^2 model. Since this model is not currently incorporated into the Ansys FLUENT flow solver, it was implemented using the user-defined function capability available in that solver.

3. Computational methodology

Flow over a two-dimensional elliptic airfoil for a fixed chord Reynolds number of 3×10^5 has been investigated in this study. This Reynolds number was chosen since it lies in the range wherein laminar-to-turbulent boundary layer transition plays a predominant role in determining aerodynamic characteristics. Since both Re and thickness ratio of the airfoil are fixed, a single geometry and mesh for elliptic airfoil was used for all CFD simulations. The surface geometry of the ellipse was defined and the grid was generated using Ansys GAMBIT software.

3.1 Geometry, grid and boundary conditions

The ellipse defining the airfoil surface was oriented in the x-y plane with a unit chord length (c = 1 m) along the positive x-axis and a maximum thickness of 0.16 units in the y-axis located at one half chord length. The upstream, downstream, top, and bottom boundaries were placed at a distance of 10 chord lengths from the ellipse. To ensure that the boundary location did not influence the flow, additional simulations were carried out in which the boundaries were placed at a distance of 20 chord lengths from the ellipse. No significant differences were seen in the results from the different geometries.

To maximize simulation efficiency, a hybrid unstructured grid topology was used. This approach allows the grid to be constructed in such a way that regions of high curvature and large flow gradient can maintain higher point densities. Using these concepts, a structurd O-type grid was generated near the airfoil surface while an unstructured triangular mesh was used for the farfield regions. A total of 480 grid points were placed on airfoil surface in such a way that more points were clustered in areas of high curvature, near the leading and trailing edges of the airfoil. The density of grid points was stretched vertically from the airfoil surface, and in the unstructured region the mesh size was decreased gradually towards the outer boundaries of the domain. In order to resolve vortex structures in the wake region, a relatively fine grid was maintained downstream of the airfoil.

A mesh boundary layer with a total depth of 0.048 chord units was used. The first point in the viscous layer was placed at a distance of 5e-05 chord units from the wall and thereby increased uniformly upto 48 point layers with a growth factor of 1.1. This spacing corresponded to a y plus value less than unity over the entire airfoil surface. Fig. 6 shows a closeup of the mesh in the vicinity of the airfoil and very near the leading edge. A total of 180,000 cells were used and the same unstructured grid was used for simulation of all turbulence models. A grid resolution study was performed by systematically refining the grid in the region of the airfoil, including both structured and unstructured regions. No considerable changes were seen in the results for meshes larger than 180,000 cells, therefore the baseline mesh was deemed to be grid independent.

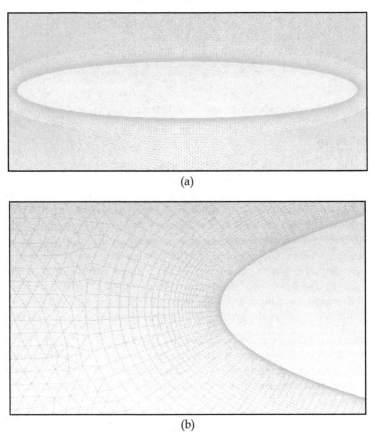

(a)

(b)

Fig. 6. Closeup of mesh in the vicinity of the elliptic airfoil (a) and closeup of mesh near the leading edge (b).

The overall domain, grid, and boundary conditions are shown in Fig. 7. The left and bottom sides of the rectangular domain were specified as velocity inlets, the right and top sides as pressure outlets and the elliptic airfoil surface as a wall. Specified inlet boundary conditions included a freestream velocity (U_∞) of 4.3822 m/s, turbulent viscosity ratio of 10 and turbulent intensity of 0.12%. Constant air density of 1.225 kg/m^3 and viscosity of 1.7894×10^{-5} kg/m-s were specified as fluid properties.

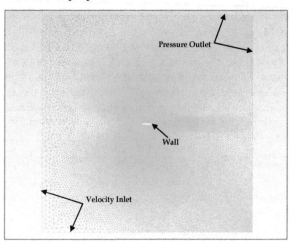

Fig. 7. Computational domain showing overall grid resolution level and boundary conditions.

3.2 Numerical setup

For a steady-state computation, the problem is said to obtain a state of convergence when the solution does not change with additional iterations, while in an unsteady computation, it must be ensured that the solution at each time step is fully converged and time-averaged flow parameters do not change with additional time steps. Simulations with with the Spalart-Allmaras (SA) fully turbulent model obtained steady-state results. For all other fully turbulent and transition-sensitive models, it was necessary to adopt an unsteady RANS (URANS) approach. A fixed time stepping method with a time step size of 0.001T was used for all unsteady simulations with a maximum of 20 iterations per time step, where T is equal to the chord length divided by the freestream velocity (T = c/U_∞). A time-step study was performed, and results obtained with a time step of 0.0005T showed no appreciable change. Unless stated otherwise, all results presented below represent time-averaged quantities.

The simulations used the pressure-based solver in Ansys FLUENT. Convective terms for all equations were discretized using a second-order upwind scheme and unsteady terms were discretized using a second-order implicit scheme for transient simulations. Upwind schemes are generally preferred for spatial discretization in order to obtain accurate results and numerical stability at high Re for incompressible flows (Nair & Sengupta, 1997). The SIMPLE scheme was used for pressure-velocity coupling, and the PRESTO scheme was used for discretization of the pressure terms. Gradients were computed using a Green-Gauss cell based method. Typically, around 6000 time steps were required for the transition-sensitive models and SST k-ω-v^2, and around 4000 time steps were required for the SST k-ω model to obtain convergence of the time-averaged quantities of flow variables. Around 5000 iterations were required to obtain a converged steady state result for the SA model.

3.3 Turbulence models

One focus of this study is the evaluation of predictive capability of transition-sensitive and curvature-sensitive RANS turbulence models versus traditional eddy-viscosity models for static elliptic airfoils at relatively low Re. Since the flow considered here is in the transitional range, there is a possibility of completely laminar, turbulent or transition from laminar-to-turbulent flow on both suction and pressure surfaces of the airfoil.

The fully turbulent (standard) eddy-viscosity models used for this study include:

- 1-equation SA model (Spalart and Allmaras, 1992)
- 2-equation SST k-ω model (Menter, 1994)

The transition-sensitive eddy-viscosity models used for this study include:

- 3-equation k-k_L-ω model (Walters and Cokljat, 2008)
- 4-equation Transition SST model (Menter et al., 2004)

The curvature-sensitive eddy-viscosity model used for this study is:

- 3-equation SST k-ω-v^2 model (Dhakal and Walters, 2011)

Each of the first four models listed above are available options in Ansys FLUENT. The curvature-sensitive model was implemented into FLUENT by the authors using User-Defined Function (UDF) subroutines.

4. Results

4.1 Airfoil Surface pressure distribution

Fig. 8 shows pressure coefficient profiles for the three fully turbulent models in comparison with experimental results. For $\alpha = 0°$, all three fully turbulent models; SA, SST k-ω and SST k-ω-v^2, predicted similar results over the suction and pressure surfaces except near the trailing edge, where the SA model predicted a slightly higher pressure than the other two models. This difference is due to the prediction of different flow separation patterns by these models near trailing edge of airfoil. Experimental values for both tripped and smooth cases predicted lower pressure distributions over the surface of airfoil in comparison with fully turbulent models. The experimental data for the tripped case showed higher pressures near the trailing edge in comparison with the smooth case. Not surprisingly, the fully turbulent models showed better agreement with the tripped case in this region, since the tripped boundary layer more closely approximates a boundary layer that is turbulent from the leading edge onward.

As angle of attack was increased, flow velocity increased near the leading edge on the suction surface, causing a sharp decrease of the pressure distribution in that region. Thereafter the pressure gradually increased over the surface as flow approached the trailing edge. Flow velocity also increased over the downstream half of the pressure side causing a decrease of pressure in that region as well. All three of the fully turbulent models predicted similar results over the suction and pressure surfaces of the airfoil. The pressure

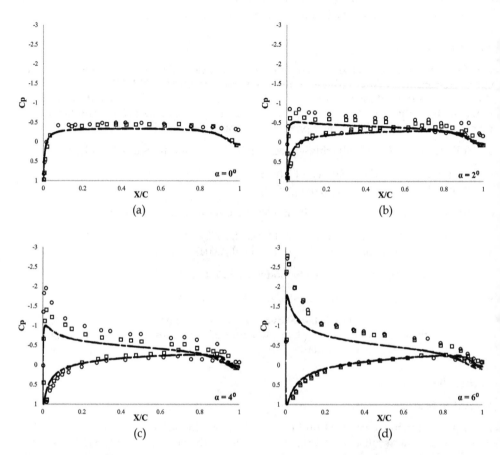

Fig. 8. Pressure coefficient profiles for fully turbulent model cases: ○ Experiment, smooth case (Kwon & Park, 2005); □ Experiment, tripped case; ▪▪▪ SST k-ω; ------ SA; —— SST k-ω-v^2.

distribution on the lower surface of the airfoil using the SA, SST k-ω and SST k-ω-v^2 models matched well with both smooth and trip case experimental values. Computational results for the suction surface showed overall reasonable agreement with trip case results, but smooth case experimental data show a much lower pressure distribution over the suction surface in comparison with the results of fully turbulent models.

Predicted pressure distributions on the airfoil surface for transition-sensitive models in comparison with experimental and fully turbulent SST k-ω results are shown in Fig. 9. For $\alpha = 0^\circ$, the k-k_L-ω and transition-sensitive SST models predicted similar results over both surfaces except in the region near the trailing edge where flow separates from the suction surface. The transition SST model predicted a higher pressure than the k-k_L-ω model near trailing edge. Differences in pressure distributions between the SST k-ω model and both transition-sensitive models for $\alpha = 0^\circ$ can be observed in Fig. 9 (a) on the downstream half of

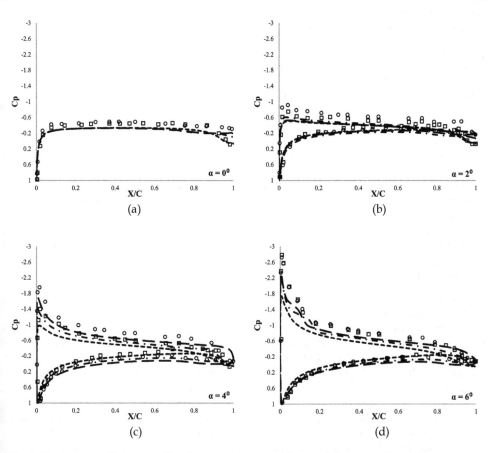

Fig. 9. Pressure coefficient profiles for transition-sensitive model cases: ○ Experiment, smooth case (Kwon & Park, 2005); □ Experiment, tripped case; ▪▪▪▪ SST k-ω; ------- Transition SST; — — k-k_L-ω.

the surface of airfoil. It appears that the k-k_L-ω model agrees more closely with the smooth case data while the transition SST model agrees with the tripped case data, although the reasons for this are not clear.

For $\alpha = 2°$, interestingly, pressure distributions for the SST k-ω and k-k_L-ω models were similar on the upstream half of the suction surface while further downstream, the SST k-ω model predicted higher pressure than k-k_L-ω. The transition SST model produced the best results for this case, particularly near the leading edge. As α increased, the k-k_L-ω model predicted better results over both the suction and pressure surfaces than the transition SST model, and compared well with the smooth case experimental data. As seen below, the lift coefficients predicted by the k-k_L-ω model were higher than those predicted by the transition SST model and this difference is attributed to the differences in surface pressure distributions. The k-k_L-ω model predicted significantly lower pressure on the suction surface and higher pressure on the bottom surface when compared to the pressure distributions of the transition SST and SST k-ω models.

While surface pressure distributions on the suction surface varied significantly as angle of attack was increased, similar changes were not observed on the pressure surface. At relatively small α, pressure coefficient profiles on the suction surface were found to rapidly reach their negative peaks near the leading edge and thereafter recover gradually on the downstream portion of the airfoil. At $\alpha = 6^0$, a slight pressure plateau region was found to exist in the experimental data next to a negative pressure peak point, followed by a sudden increase in pressure coefficient next to the plateau region. This characteristic pressure distribution is indicative of formation of a laminar separation bubble near the leading edge (Hu & Yang, 2008). Although a tiny separation bubble was observed in velocity vector plots of SST k-ω, unlike pressure plots of transition-sensitive models, no significant changes were observed in pressure distributions of SST k-ω over suction surface due to presence of separation bubble. This must be due to the size of separation bubble produced in that region. As first explained by Tani (1964) and later reviewed by Shyy et al. (1999) and Lian & Shyy (2007), long separation bubbles generally cover considerable portion of airfoil surface and affect inviscid pressure and velocity distributions around the airfoil, whereas, short bubbles cover small portion of surface and do not affect pressure and velocity distributions.

Both k-k_L-ω and Transition SST models captured the laminar separation bubble formed on suction surface near leading edge of airfoil at $\alpha = 6^0$. The separation bubble stayed on suction surface for a large range of α prior to stall. As α increased, the laminar separation bubble moved towards the leading edge and the size of the bubble reduced gradually. The LSB formed on suction surface can be characterized by a theoretic model given by Russell (Russell, 1979). According to his model, flow separation, transition and reattachment locations on suction surface can be determined by the pressure distribution over the surface of airfoil. The point from where laminar boundary layer separation occurs on airfoil surface refers to separation point. The separated boundary layer undergoes transition to turbulence due to amplification of flow instabilities at transition point and reattaches to airfoil surface as a turbulent boundary layer at reattachment point. The separation bubble formed on a low Re airfoil surface generally includes a laminar and turbulent portion. Distance between separation and transition point is laminar portion and distance between transition and reattachment point is the turbulent portion of bubble (Horton, 1967). At angles of attack greater than 13^0 for transition SST and 16^0 for k-k_L-ω, a negative pressure peak near the leading edge was found to decrease and the pressure plateau region became nonexistent. Also, surface pressures on the downstream portion of the airfoil for both suction and pressure sides remained nearly constant. This pressure distribution indicates that the airfoil reached stall at that point.

4.2 Velocity distribution around the airfoil

At $\alpha = 0^0$, the flow over the airfoil was symmetrical and flow separation occurred near the trailing edge. Two attached symmetrical vortices with opposite rotation developed aft of airfoil. As angle of attack was increased, the flow separation point on the suction surface moved towards the leading edge, while the flow separation point on the pressure surface moved towards the trailing edge causing an asymmetric flow around airfoil. The two vortices aft of the airfoil moved upwards causing the upper vortex to be larger than lower vortex. Figs. 10, 11 and 12 show velocity vector plots for SST k-ω at $\alpha = 0^0$, 7^0 and 12^0 respectively. From Fig. 10, two symmetrical vortices created near the trailing edge of airfoil can be observed. As angle of attack increases, the cross section area of lower vortex gradually decreases and the vortex itself shifts towards the suction side of airfoil. The shift in the locations of vortices near trailing edge can be observed from Figs. 11 and 12. In addition, the leading edge stagnation point moved downwards to pressure side as α increased.

Interestingly, a tiny separation bubble was observed to appear for the fully turbulent SST k-ω and SST k-ω-v^2 models at about $\alpha = 11^0$. This bubble stayed near leading edge of the suction surface and finally burst out at $\alpha = 18^0$ for SST k-ω and $\alpha = 13^0$ for SST k-ω-v^2, causing reverse flow over entire suction surface, indicating that airfoil had stalled. No separation bubbles were observed in simulation results for the SA model.

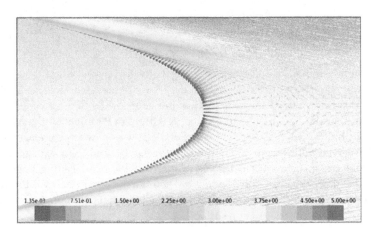

Fig. 10. Velocity vectors near the trailing edge for SST k-ω at $\alpha = 0^0$.

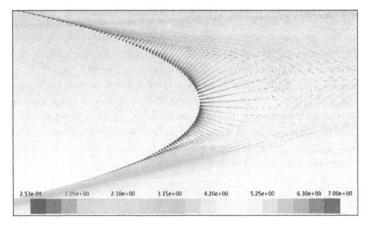

Fig. 11. Velocity vectors near the trailing edge for SST k-ω at $\alpha = 7^0$.

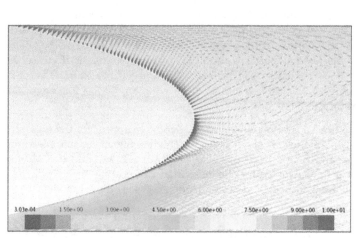

Fig. 12. Velocity vectors near the trailing edge for SST k-ω at $\alpha = 12^{\circ}$.

For all three fully turbulent models, flow remained attached over most of the airfoil surface and separated at about 95% of chord length at $\alpha = 0^{\circ}$. As α increased, flow separation points on the suction surface moved towards the leading edge, while flow separation points on pressure surface moved towards the trailing edge. Although for initial angles of attack, flow separated from trailing edge earlier in SST k-ω and SST k-ω-v^2 when compared with SA, a change was observed in flow separation locations for SST k-ω and SST k-ω-v^2 at $\alpha = 11^{\circ}$. This is likely due to the formation of a tiny separation bubble over leading edge of suction surface mentioned above. The bubble served to keep the flow attached over the surface but as adverse pressure gradient became more severe, the bubble finally burst out resulting in a complete reverse flow over the suction surface.

Flow behavior predicted by the transition-sensitive models was markedly different from that observed for the fully turbulent models. Initially, flow over most of the airfoil surface was laminar for $\alpha = 0^{0}$ to 5^{0} and hence, flow separated earlier from the trailing edge when compared to the flow separation points predicted by the fully turbulent cases. At a flow angle of 6^{0}, the adverse pressure gradient became severe enough that the laminar boundary layer separated, transitioned to turbulent flow and reattached to the suction surface near the leading edge. This flow behavior was captured accurately by both the k-k_L-ω and transition SST models. The reattached turbulent boundary layer was reenergized and hence separated from the suction surface at a farther location significantly farther downstream on the suction surface. As α continued to increase, the separation bubble moved towards the leading edge of the airfoil and the bubble size reduced gradually. As adverse pressure gradient became more severe, the separation bubble burst, resulting in flow reversal over entire suction side of airfoil indicating that the airfoil had stalled.

4.3 Lift and drag coefficients of the airfoil

The lift coefficient (c_l) and drag coefficient (c_d) plots for the fully turbulent model results are shown in Fig. 13 and compared with smooth and tripped case experimental results

(Kwon & Park, 2005). It is observed that all the three fully turbulent models failed to capture the flow transition behavior over the airfoil and hence, a discrepancy in lift coefficient is observed in the CFD results in comparison with experimental results for all values of α prior to stall. Both the SA and SST k-ω models predicted stall at $\alpha = 17^0$ and 16^0 respectively, which is considerably later than the experimental data indicate. Interestingly, the SST k-ω-v^2 model predicted a stall point close to experimental results at $\alpha = 12^0$, although lift values prior to stall were still not accurately predicted. However, all three fully turbulent models accurately predicted the drag coefficient values in comparison with experimental results. Discrepancies in c_d values for the SA and SST k-ω models are only observed at angles of attack greater than 10^0. This is primarily due to the delayed prediction of airfoil stall.

Lift and drag coefficient plots for both transition-sensitive models are shown in Fig. 14 and compared with experimental data and with SST k-ω results. It was observed previously that the transition sensitive models accurately predict the flow transition behavior, yielding laminar boundary layers up to 6^0 and separation bubbles near the leading edge for $\alpha > 6^0$. As a consequence, the slope of the lift coefficient for $\alpha < 6^0$ was greater than for the fully turbulent models and in better agreement with the experimental data. The formation of the laminar separation bubble caused a shift in the lift curve slope between $\alpha = 4^0$ and 6^0, and the lift curve slope approximately matched the fully turbulent models beyond that point. Although flow transition behavior was captured accurately by both k-k_L-ω and transition SST models, they notably failed to predict the airfoil stall point, although the transition SST model yields a closer result than the k-k_L-ω model. The c_d values of both transition-sensitive models compared relatively well with experimental results prior to the stall point. Observing Figs. 13 and 14, the results of this study seem to suggest that accurate prediction of aerodynamic characteristics using linear eddy-viscosity RANS models can be best achieved by a combination of transition-sensitive modeling, which is necessary to predict increased lift values prior to stall, and curvature-sensitive modeling, which is necessary to correctly resolve the stall point.

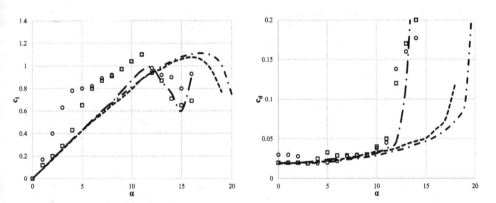

Fig. 13. Lift and drag coefficient curves for fully turbulent models: ○ Experiment, smooth case (Kwon & Park, 2005); □ Experiment, tripped case; ▪▪▪▪ SST k-ω; ------ SA; —·— SST k-ω-v^2.

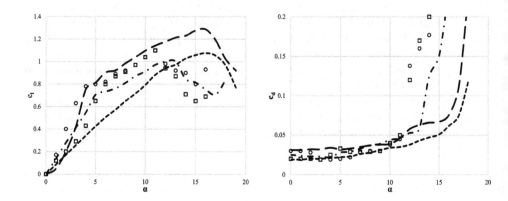

Fig. 14. Lift and drag coefficient curves for transition-sensitive models: ○ Experiment, smooth case (Kwon & Park, 2005); □ Experiment, tripped case; •••• SST k-ω; ······ Transition SST; — — k-k_L-ω.

5. Conclusions

Numerical simulations were performed for flow over an elliptic airfoil at varying angles of attack and at a low/transitional Reynolds number of 3×10^5. Simulations were carried out using relatively new transition- and curvature-sensitive eddy viscosity RANS models as well as traditional fully turbulent eddy viscosity models. CFD results were compared to the wind tunnel experimental test results of Kwon & Park (2005). Results indicate that the transition-sensitive models are indeed necessary to accurately predict the transition and separation flow behavior on the suction surface of the airfoil. Both transition-sensitive models — k-k_L-ω and transition SST — reproduce the separation bubble that appears near the leading edge of the airfoil over a relatively large range of angle of attack prior to stall. Consequently, the lift characteristics were better predicted, showing elevated levels and nonlinear increase similar to the experimental data. In contrast, the fully turbulent models — SA, SST k-ω and curvature-sensitive SST k-ω-v^2 — failed to accurately predict the boundary layer separation and reattachment phenomena and therefore showed large discrepancies in lift prediction. Although the SST k-ω and SST k-ω-v^2 models predicted a small separation bubble near the leading edge at $\alpha = 11°$, the effect of the separation bubble on pressure and velocity distributions over the airfoil surface was minimal. Interestingly, both the transition-sensitive models as well as the SA and SST k-ω fully turbulent models failed to accurately predict the stall point. However, the curvature sensitive SST k-ω-v^2 successfully predicted the stall point very close to experimental results. These results indicate that accurate RANS prediction of aerodynamic characteristics both pre- and post-stall require models that respond correctly to transitional as well as curvature effects. While a comprehensive turbulence model considering both curvature and flow transition effects at low Reynolds numbers is not yet available, future research efforts will seek to integrate the existing models used here to improve overall predictive capability for these problems.

6. References

Assel, T. W. (2007). Computational study of flow over elliptic airfoils for rotor/wing unmanned aerial vehicle applications, Thesis/Dissertation, University of Missouri – Rolla

Carmichael, B.H. (1981). Low Reynolds Number Airfoil Survey. NASA CR – 165803, Vol. 1

Choudhury, D. (1993). Introduction to the renormalization group method and turbulence modeling, Fluent Inc. Technical Memorandum, TM-107

Dhakal, T. P. & Walters, D. K. (2011). A Three-Equation Variant of the SST k-ω Model Sensitized to Rotation and Curvature Effects. J Fluid Eng - T ASME, Vol. 133, No. 11, (November 2011), pp. 11201:1-9

Diwan, S. S. & Ramesh, O. N. (2007). Laminar separation bubbles: Dynamics and Control. Sadhana, Vol. 32, Parts 1 & 2, (April 2007), pp. 103-109

Durbin, P. (2011). Review: Adapting Scalar Turbulence Closure Models for Rotation and Curvature. J Fluid Eng - T ASME, Vol. 133, (June 2011), pp. 061205:1-8

Gaster, M. (1967). The Structure and Behavior of Separation Bubbles. Reports and Memoranda No. 3595. pp. 1-9

Genc, M. S.; Kaynak, U. & Lock, G. D. (2009). Flow over an aerofoil without and with a leading-edge slat at a transitional Reynolds number, Proceedings of the Institutions of Mechanical Engineers, Part G: Journal of Aerospace Engineering, (March 2009), Vol. 223, pp. 217-231

Genc, M. S.; Kaynak, U. & Yapici, H. (2011). Performance of Transition Model for Predicting Low Re Aerofoil Flows without/with Single and Simultaneous Blowing and Suction. European Journal of Mechanics B/Fluids, pp. 218-235

Horton, H. P. (1967). A Semi-empirical Theory for the Growth and Bursting of Laminar Separation Bubbles. PhD thesis, Queen Mary College, University of London

Hu, H. & Yang, Z. (2008). An Experimental Study of the Laminar Flow Separation on a Low-Reynolds–Number Airfoil. J Fluid Eng - T ASME, Vol. 130, (May 2008), pp. 1-11

Jahanmiri, M. (2011). Laminar Separation Bubble: Its Structure, Dynamics and Control. Research Report, ISSN 1652-8549

Johnson, S. A., Thompson, M. C. & Hourigan, K. (2001). Flow Past Elliptical Cylinders at Low Reynolds Numbers, Proceedings of AFMS 2001 14th Australasian Fluid Mechanics Conference, Adelaide University, Adelaide, Australia, December 10-14, 2001

Jones, B. M. (1938). Stalling. Journal of the Royal Aeronautical Society, Vol. 38, pp. 747-770

Kim, M. S. & Sengupta, A. (2005). Unsteady Viscous Flow over Elliptic Cylinders At Various Thickness with Different Reynolds Numbers. Journal of Mechanical Science and Technology, Vol. 19, No. 3, (December 2005), pp. 877-886

Kwon, K. & Park, S. O. (2005). Aerodynamic Characteristics of an Elliptic Airfoil at Low Reynolds Number. Journal of Aircraft, Vol. 42, No. 6, (December 2005), pp. 1642-1644

Kwon, K.; Chang, B.; Lee, J. & Park, S. O. (2006). Boundary Layer Transition Measurement over an Airfoil by Using PIV with High Magnification. Lisbon, Portugal, pp. 1-8

Lian, Y. & Shyy, W. (2007). Laminar-Turbulent Transition of a Low Reynolds Number Rigid or Flexible Airfoil. AIAA Journal, Vol. 45, No. 7, (July 2007), pp. 1501-1513, ISSN 0001-1452

Lin, J. C. M. & Pauley, L. L. (1996). Low-Reynolds number separation on an airfoil. AIAA Journal, Vol. 34, No. 8, pp. 1570-1577.

Lissaman, P. B. S. (1983). Low Reynolds Number Airfoils. *Annual Review Fluid Mechanics*, No. 15, pp. 223-239

Menter, F. (1994). Two-equation eddy viscosity turbulence models for engineering applications. *AIAA Journal*, Vol. 32, No. 8, pp. 1598-1605, ISSN 0001-1452

Menter, F. R.; Langtry, S. R.; Likki, Y. B.; Suzen, P. G.; Huang, S. & Volker. (2004) A Correlation Based Transition Model Using Local Variables. *Proceedings of ASME Turbo Expo 2004*, ISBN GT2004-53452, Vienna, Austria, pp. 57-67

Nair, M. T. & Sengupta, T. K. (1997). Unsteady Flow Past Elliptical Cylinders. *Journal of Fluids and Structures*, Vol. 11, No. 6, (August 1997), pp. 555-595

Newcome, L. R. (2004). *Unmanned Aviation: A Brief History of Unmanned Aerial Vehicles*, American Institute of Aeronautics and Astronautics (AIAA), ISBN 1-56347-644-4, Virginia, USA

Pandya, S. A. & Aftosmis, M. J. (2001). Computation of External Aerodynamics for a Canard Rotor/Wing Aircraft, *Proceedings of AIAA 2001 39th AIAA Aerospace Sciences Meeting and Exhibit*, Reno, Nevada, USA, January 8-11, 2001

Russell, J. (1979). Length and Bursting of Separation Bubbles: A Physical Interpretation, *Proceedings of Science and Technology of Low Speed Motorless Flight NASA Conference*

Rutherford, J. W.; Bass, S.M. & Larsen, S. D. (1993). Canard Rotor/Wing: A Revolutionary High-Speed Rotorcraft Concept. *AIAA paper 93-1175*, (February 1993)

Sandham, N. D. (2008). Transitional Separation Bubbles and Unsteady Aspects of Aerofoil Stall. *The Aeronautical Journal*, Vol. 112, No. 1133, (July 2008), pp. 395-404

Schubauer, G. B. Air Flow in Boundary Layer of an Elliptic Cylinder. *National Advisory Committee for Aeronautics*, Report No. 652, pp. 207-226

Shyy, W.; Berg, M. & Ljungqvist, D. (1999). Flapping and Flexible Wings for Biological and Micro Vehicles. *Progress in Aerospace Sciences*, Vol. 35, No. 5, pp. 455-506

Spalart, P. R. & Allmaras, S. R. (1992) A One-Equation Turbulence Model for Aerodynamic Flows. AIAA Paper No. 92-0439.

Tani, I. (1964). Low Speed Flows Involving Bubble Separation. *Progress in Aerospace Sciences*, Vol. 5, pp. 70-103

Walters, D. K. & Cokljat, D. (2008). A Three-Equation Eddy-Viscosity Model for Reynolds-Averaged Navier-Stokes Simulations of Transitional Flow. *J Fluid Eng - T ASME*, Vol. 130, (December 2008), pp. 121401:1-14

Walters, D. K. & Leylek, J. H. (2004) A New Model for Boundary Layer Transition Using a Single-Point RANS Approach. *Journal of Turbomachinery*, Vol. 126, No. 1, (January 2004), pp. 193-202, ISSN 0889-504X

Walters, D. K. Physical Interpretation of Transition-Sensitive RANS Models Employing the Laminar Kinetic Energy Concept. *ERCOFTAC Bulletin*, Vol. 80, pp. 67-71.

Yang, Z. & Hu, H. (2008). Laminar Flow Separation and Transition on a Low-Reynolds-Number Airfoil. *Journal of Aircraft*, Vol. 45, No. 3, (June 2008), pp. 1067-1070

Yarusevych, S.; Sullivan, P. E. & Kawall, J. G. (2006). Coherent Structures in an Airfoil Boundary Layer and Wake at Low Reynolds Numbers. *American Institute of Physics*, (April 2006), Vol. 18, pp. 1-11

Zahm, A.F.; Smith, R.H. & Louden, F. A. Forces on Elliptic Cylinders in Uniform Air Stream. *National Advisory Committee for Aeronautics*, Report No. 289, pp. 215-232

Part 3

Flow Control

5

A Methodology Based on Experimental Investigation of a DBD-Plasma Actuated Cylinder Wake for Flow Control

Kelly Cohen[1], Selin Aradag[2], Stefan Siegel[3],
Jurgen Seidel[3] and Tom McLaughlin[3]
[1]*University of Cincinnati, Ohio,*
[2]*TOBB University of Economics and Technology*
[3]*US Air Force Academy, Colorado*
[1,3]*USA*
[2]*Turkey*

1. Introduction

The main purpose of flow control is to improve the mission performance of air vehicles. Flow control can either be passive or active and active flow control is further characterized by open-loop or closed-loop techniques. Gad-el-Hak (1996) provides an insight into the advances in the field of flow control. Research of closed-loop flow control methods has increased over the past two decades. Cattafesta et al (2003) provide a useful classification of active flow control.

Before proceeding into the details of modeling and control, it is imperative to appreciate the reasons as to why closed-loop control is of importance and the main advantages associated with its application to flow control problems. It is advantageous to opt for closed-loop flow control for the following reasons:

1. It enables addressing problems that have over the years not been solved using passive means and /or open-loop techniques.
2. It provides performance augmentation of an open-loop flow control system.
3. It lowers the amount of energy required to manipulate the flow to induce the desired behavior. This aspect affects actuation requirements and may be a deciding factor for the feasibility of implementation.
4. It enables adaptability to a wider operating envelope, thereby limiting the drop in performance associated with multiple design working points.
5. It provides design flexibility and robustness.

Several applications of closed-loop control have been reported in literature, namely, specific areas of interest include flow-induced cavity resonance. (Cattafesta et al, 2003, Samimy et al, 2003), vectoring control of a turbulent jet (Rapoport et al, 2003), separation control of the NACA-4412 Airfoil (Glauser, 2004) and control of vortex shedding in circular cylinder wakes (Gerhard et al, 2003, Gillies, 1995). The ability to control the wake of a bluff body could be used to reduce drag, increase mixing and heat transfer, and vibration reduction.

We can consider the cylinder wake problem. In a two-dimensional cylinder wake, self-excited oscillations in the form of periodic shedding of vortices referred to as the von Kármán Vortex Street. Shedding of counter-rotating vortices is observed in the wake of a two-dimensional cylinder above a critical Reynolds number (Re ~ 47, non-dimensionalized with respect to free stream speed and cylinder diameter). An effective way of suppressing the self-excited flow oscillations, without making changes to the geometry or introducing vast amounts of energy, is by the incorporation of active closed-loop flow control (Gillies, 1995). A closed-loop flow control system is comprised of a controller that introduces a perturbation into the flow, via a set of actuators, to obtain desired performance. Furthermore, the controller acts upon information provided by a set of sensors. During the past years, the closed-loop flow control program research effort at the United States Air Force Academy (USAFA) focused on developing a suite of low-dimensional flow control tools based on the low Reynolds numbers (Re ~ 100-300) cylinder wake benchmark (Cohen et al, 2003, Cohen et al, 2004, Cohen et al, 2005, Cohen et al, 2006a, Siegel et al, 2003a) Several computations and experiments were also performed for the cylinder wake at high Reynolds numbers (Re=20000) (Aradag, 2009, Aradag et al, 2010)

Energy is introduced into the flow via actuators and the flow field in the wake of a cylinder may be influenced using several different forcing techniques with the wake response being similar for different types of forcing (Gillies, 1998) The following forcing methods have been employed: external acoustic excitation of the wake, longitudinal, lateral or rotational vibration of the cylinder, and alternate blowing and suction at the separation points (Gillies, 1998). Work at USAFA has shown that the Dielectric Barrier Discharge (DBD) plasma actuator (Munska and McLaughlin, 2005) is an effective means of forcing at higher frequencies without mechanical movement. This relatively simple actuation device is composed of two thin electrodes separated by a dielectric barrier. When an AC voltage is applied to the electrodes, a plasma discharge propagates from the edge of the exposed electrode over the insulated electrode. The emergence of this plasma is accompanied by a coupling of directed momentum into the surrounding air as the plasma propagates over the buried electrode during each oscillation forcing cycle (Enloe et al, 2004). This momentum can effectively alter a moving flow or generate flow in the direction of plasma propagation, as several application-based papers have shown (List et al, 2003, Asghar and Jumper, 2003, Bevan et al, 2003). The non-mechanical nature of the plasma actuator makes it ideal for high Re flow control applications. Its high fundamental operating frequency suggests it can be effective over a very wide bandwidth (by fluid time scale standards). This enables operation over a much broader range of frequencies than mechanical actuators. It has no moving parts, and has no resonant frequency. Munska and McLaughlin (2005) established that plasma actuators can achieve vortex shedding lock-in and span-wise coherence over a range of forcing conditions. They employed a cylinder arrangement similar to that of Asghar and Jumper (2003), with electrodes at ±90° and Re up to 88x10³, and used a similar amplitude-modulated forcing scheme.Low-dimensional modeling is a vital building block when it comes to realizing a structured model-based closed-loop flow control strategy. For control purposes, a practical procedure is needed to represent the velocity field, governed by the Navier Stokes partial differential equations, by separating space and time. A common method used to substantially reduce the order of the model is Proper Orthogonal Decomposition (POD). This method is an optimal approach in that it will capture a larger amount of the flow energy in the fewest modes of any decomposition of the flow. The two dimensional POD method was used to identify the characteristic features, or modes, of a cylinder wake as demonstrated by Gillies (1998) and Gerhard et al (2003).

The major building blocks of the structured approach presented here are comprised of a reduced-order POD model, a state estimator and a controller. The desired POD model contains an adequate number of modes to enable accurate modeling of the temporal and spatial characteristics of the large scale coherent structures inherent in the flow in order to model the dynamics of the flow. A Galerkin projection may be used to derive a set of reduced order ordinary differential equations by projecting the Navier-Stokes equations on to the modes (Holmes et al, 1996). Further details of the POD method may be found in Holmes et al (1996). A common approach referred to as the method of "snapshots" introduced by Sirovich (1987) is employed to generate the basis functions of the POD spatial modes from flow-field information obtained using either experiments or numerical simulations. This approach to controlling the global wake behavior behind a circular cylinder was effectively employed by Gillies (1998) and Noack et al (2004) and is also the approach followed in the current research effort.

For practical applications, it is important to estimate the state of the flow, i.e. the relevant POD time coefficients, using body mounted sensors. The advantages of body mounted sensors are:

1. Simple, relatively inexpensive and reliable.
2. Essential for real-life, closed-loop flow control applications where the direct measurement of the separated wake flow field is cumbersome (if not impossible)
3. Enable collocation of sensors and actuators, which eliminates substantial phase delays effecting controller design.

Pressure sensors, mounted on the surface have been used on a back-ward facing ramp by Taylor and Glauser (2004) and by Glauser et al (2004) on a NACA 4412 airfoil. Recent efforts have successfully demonstrated estimation of the time coefficients of the POD model for a "D" shaped cylinder for laminar flow at low Reynolds numbers (Cohen et al, 2004, Stalnov et al, 2005). The body mounted sensors may measure skin friction or surface pressures, as is done in this effort. The intention of the proposed strategy is that the measurements, provided by a certain configuration of body mounted pressure sensors placed on the model surface, are processed by an estimator to provide the real-time estimates of the POD time coefficients that are used to close the feedback loop. The estimation scheme is to behave as a modal filter that "combs out" the higher modes. The main aim of this approach is to thereby circumvent the destabilizing effects of observation "spillover". The estimation scheme may be based on the linear stochastic estimation procedure introduced by Adrian et al (1977) or a quadratic stochastic estimation proposed by Murray and Ukeiley (2002) as well as by Ausseur et al (2006).

This chapter is organized as follows: The following section provides the main objective of this chapter. The basic approach to feedback flow control for turbulent wake flows is presented in Section III. A wind tunnel experiment of a plasma actuated cylinder wake, at a Reynolds number of 20,000, is described in Section IV. Preliminary experimental results using POD and a Neural Network based estimator and a subsequent discussion are presented in Section V. Finally, the conclusions of this research effort and recommendations for future work are summarized in Section VI.

2. Aims and concerns

Technological advances in sensors, actuators, on-board computational capability, modeling and control sciences have offered a possibility of seriously considering closed-loop flow control for practical applications. The main strategies to closed-loop control are a model-independent, full-order optimal control approach based on the Navier-Stokes equations and a reduced order model strategy. This effort emphasizes the methodology based on the low-dimensional, proper orthogonal decomposition method applied to the problem concerning the suppression of the von Kármán vortex-street in the wake of a circular cylinder. Focus is on the validity of the low-dimensional model, selection of the important modes that need representation, incorporation of ensembles of snapshots that reflect vital transient phenomena, selection of sensor placement and number, and linear stochastic estimation for mapping of sensor data onto modal information. Furthermore, additional issues surveyed include observability, controllability and stability of the closed-loop systems based on low-dimensional models. Case studies based on computational and experimental studies on the cylinder wake benchmark are presented to illuminate some of the important issues.

3. Research methods

3.1 Closed loop control methodology

Based on the research effort at the USAF Academy over the past years, a methodology for approaching closed-loop flow control has been developed. This approach has been applied to control of laminar bluff body wakes at low Reynolds numbers (Re~50-180). In this work, this methodology is extended to higher Reynolds number turbulent wakes (Re~20,000). A schematic representation of the setup is presented in Figure 1.

Fig. 1. Methodology for Closed-Loop Flow Control.

The following is a more detailed look into each of the six steps:

a.   Identification of the "Lock-In" Region

In order to obtain a meaningful low-order representation of the flow, it is imperative that the behavior of the flow be constrained so that it can be characterized using a relatively small number of parameters. A good example that illustrates this feature is the "lock-in" envelope of a cylinder wake. The cylinder wake flow can be forced in an open loop fashion using sinusoidal displacement of the cylinder with a given amplitude and frequency. Koopmann (1967) investigated the response of the flow to this type of forcing in a wind tunnel experiment. He found a region around the natural vortex shedding frequency where he could achieve "lock-in", which is characterized by the wake responding to the forcing by establishing a fixed phase relationship with respect to the forcing. The frequency band around the natural vortex shedding frequency for which lock-in may be achieved is amplitude dependent. In general, the larger the amplitude, the larger the frequency band for which lock-in is possible. However, there exists a minimum threshold amplitude below which the flow will not respond to the forcing any more. In Koopmann's experiment (1967), this amplitude was at 10% peak displacement of the cylinder. Siegel et al (2003a) show that for a circular cylinder, at Reynolds number of 100, a closed-loop controller operating within the "lock-in region" achieves a drag reduction of close to 90% of the vortex-induced drag, and lowers the unsteady lift force by the same amount.

Recently, the dielectric barrier discharge (DBD) plasma electrode has been developed as a flow control actuator, showing the ability to affect flow behavior in a range of applications. McLaughlin et al (2006) applied the DBD plasma actuator to a circular cylinder at Reynolds numbers of up to 3×10^5. Hot film measurements show that vortex shedding frequency can be driven to the actuator forcing frequency, within the lock-in range, at all Reynolds numbers investigated. The wake forced with plasma actuators exhibits "lock-in" behavior similar to that previously reported by Koopmann using cylinder displacement for forcing (Munska and McLaughlin, 2005).

b.   Open-Loop, Transient Excitation using Actuators

Since the intended use of the low dimensional model, based on POD, is feedback flow control, the low dimensional state of the flow field needs to be accurately estimated as input for a controller. This poses the problem of snapshot selection: For the state to which the feedback controller drives the flow, usually no snapshots are available beforehand. We investigated POD bases derived from steady state, transient startup and open loop forced data sets for the two dimensional circular cylinder wake at Re = 100. None of these bases by itself is able to represent all features of the feedback controlled flow field. However, a POD basis derived from a composite snapshot set consisting of both transient startup as well as open loop forced data accurately models the features of the feedback controlled flow. For similar numbers of modes, this POD basis, which can be derived *a* priori, represents the feedback controlled flow as well as a POD model developed from the feedback controlled data a posteriori. These findings have two important implications: Firstly, an accurate POD basis can be developed without iteration from unforced and open loop data. Secondly, it appears that the feedback controlled flow does not leave the subspace spanned by open loop and unforced startup data, which may have important implications for the performance limits of feedback flow control. Further details on this approach are presented by Siegel et al (2005a) and Seidel et al (2006).

An important aspect of the developed methodology is to obtain a low-dimensional model that can predict the modal behavior of the flow when subject to various forcing inputs within the lock-in region. The emphasis is on the robustness of the predictive capability of the model. The main aim here is to predict the time histories of the time coefficients of the truncated POD model under the influence of open-loop control within the lock-in region. For the low Reynolds number, circular cylinder wake problem, Cohen et al (2006b) used nine different data sets, as marked in Figure 2, for the open loop forced cases at 10, 15, 20, 25 and 30 percent cylinder displacement. Some of the cases use 5-10% lower or higher frequency at 30% displacement, which is still within the lock-in region. In this example, the 25 percent cylinder displacement sinusoidal forcing serves as design point for model development.

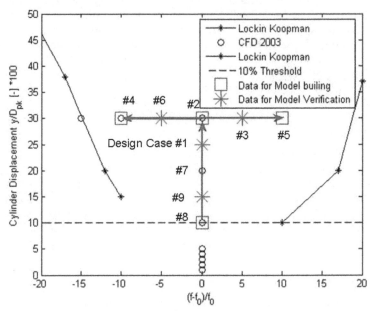

Fig. 2. Model Building within "Lock-In" Envelope for the Circular Cylinder Wake.

c.   Development of a Low-Dimensional Model (LOM) based on POD

In the developed approach, the main advantage of POD, namely its optimality and thus ability to capture the global behavior of a flow field with a minimum number of modes, is combined with established system identification techniques developed for the modeling of dynamical systems. Over the past decades, the controls community has developed methods to identify the dynamic properties of complex structures based on experimental measurements. These rely on the acquisition of transient measurements based on a known excitation input to the system. So called System Identification methods are then used to develop a dynamical mathematical model that can be used later for design and analysis of an effective control law as well as dynamic observer development. The main emphasis is then to develop an effective system identification technique that captures the dynamics of the time dependent coefficients of the POD modes with respect to transient actuation inputs within the lock-in region.

An important question that needs to be answered is: "What are the desired characteristics most sought after in a low-order, POD based model?" It is imperative to understand that given the complexity of the problem at hand, it may not be possible to address this problem with off-the-shelf methods but instead we propose a unique synthesis of software tools that appear to be promising. The important features are:

1. **Structured scalable methodology:** Developing an ad-hoc approach as demonstrated by Gillies (1995) using the least squares technique may address a particular problem for a given design point under certain conditions but is not generic enough. An approach which may be applied to a wide range of flow conditions (Reynolds numbers) is preferable. Another important principle is to let the data determine the dynamic complexity, i.e. the number of POD modes, of the reduced order model using the amount of truncated energy as a criterion. This approach differs from that of Noack et al (2003) that uses first principles to make an a priori decision on the number and nature of the modes.

2. **Numerical issues and model stability:** The non-linearity and scaling characteristics of the temporal coefficients lead to numerical stability issues which undermine the development and analysis of effective estimation/control laws. A major numerical problem which emerges using the Galerkin projection is the effect of noisy data on higher order derivatives required for model development. In order to assure model stability, the system identification community very widely uses the ARX dynamic model structure. A salient feature of the ARX model is that it is inherently stable even if the dynamic system to be modeled is unstable. This characteristic of ARX models often lends itself to successful modeling of unstable processes as described by Nelles (2001).

3. **Model validity and robustness:** An important task of model building is the determination of the dynamic envelope within which the model is valid. Then, after deriving a model, one needs to ensure that the model is capable of providing relatively accurate predictions within the region of validity. Validity and robustness are necessary conditions to design a model based observer/controller. For this reason, we validate the model by comparing it to a high resolution, closed-loop, CFD simulation.

4. **Universal approximation of non-linear mappings:** The decision was to look into universal approximators, such as artificial neural networks (ANN), for their inherent robustness and capability to approximate any non-linear function to any arbitrary degree of accuracy (Cybenko, 1989). The ANN, employed in this effort, in conjunction with the ARX model is the mechanism with which the dynamic model is developed using the POD time-coefficients extracted from the high resolution CFD simulation. Non-linear optimization techniques, based on the back propagation method, are used to minimize the difference between the exact POD time coefficients and the ANN based estimate while adjusting the weights of the model (Haykin, 1999).

The ARX-ANN algorithms used in this effort are a modification of the toolbox developed by Nørgaard et al (2000). After the POD time coefficients were extracted, a basic single hidden layer ANN-ARX architecture was selected. The training set was then developed using Input-Output data obtained from CFD simulations. The model was validated for off-design cases and if the estimation error was unacceptable, then the ANN architecture was modified. This cycle repeated until estimation errors were acceptable for all off-design cases. Cohen et al (2006b) successfully demonstrated this approach for modeling of a cylinder wake at Reynolds numbers of 100.

Most of the modeling effort has been based on the velocity field. The low Reynolds number cylinder wake flow is two dimensional in nature, simplifying the spatial characteristics of the POD modes. However, as the Reynolds number is increased to turbulent regimes or as the geometry of the model becomes more complex, the flow field becomes three dimensional in nature.

a.   Sensor Configuration and Estimator Development

A major design challenge lies in finding an appropriate number of sensors and their locations that will best enable the flow estimation. For low-dimensional control schemes to be implemented, a real-time *estimation* of the modes present in the wake is necessary, since it is not possible to measure them directly. Velocity field (Cohen et al, 2006a), surface pressure (Cohen et al, 2004), or surface skin friction (Stalnov et al, 2005) measurements, provided from either simulation or experiment, are used for estimation. This process leads to the state and measurement equations, required for design of the control system. For practical applications it is desirable to reduce the information required for estimation to the minimum. The spatial modes obtained from the POD procedure provide information that can be used to place sensor in locations where modal activity is at its highest. These areas would be the maxima and minima of the spatial modes (Cohen et al, 2006a). Placing sensors at the energetic maxima and minima of each mode is the basic hypothesis of the developed approach and the purpose of the CFD simulation is to design a sensor configuration which is later validated using experiments (Cohen et al, 2004).

The estimation scheme may be based on the linear stochastic estimation procedure introduced by Adrian (1977) a quadratic stochastic estimation (Ausseur et al, 2006) or in the form of an artificial neural network estimator, ANNE (Cohen et al, 2006c). Cohen et al (2006c) compare the effectiveness of the conventional LSE versus the newly proposed ANNE. The development of the procedure was based on CFD simulations of a cylinder at a Reynolds number of 100. Results show that for the estimation of the first four modes, it is seen that for the design condition (no noise) 4 sensors using ANNE provide significantly better results than 4 sensors using LSE. For the estimation of the first four modes, we show that a considerably smaller number of sensors using ANNE estimation provide better results than more sensors using LSE estimation. Furthermore, ANNE displays robust behavior when the signal to noise ratio of the sensors is artificially degraded.

b.   Development and Analysis of a Control Law

A simple approach to control the von Kármán Vortex Street behind a two dimensional circular cylinder, based on the proportional feedback of the estimate of *just* the first POD mode was presented by Cohen et al (2003). A stability analysis of this control law was conducted after linearization about the desired equilibrium point and conditions for controllability and asymptotic stability were developed. The control approach, applied to the 4 mode cylinder wake POD model at a Reynolds number of 100 stabilizes the POD based low-dimensional wake model. While the controller uses only the estimated amplitude of the first mode, all four modes are stabilized. This suggests that the higher order modes are caused by a secondary instability. Thus they are suppressed once the primary instability is controlled. This simple control approach was later modified by Siegel et al (2003a) when applying it to a high resolution CFD simulation. An adaptive gain strategy, based on the estimation of the "mean-flow" mode incorporated to tune the phase of a Proportional-

Differential (PD) controller was used (Siegel et al, 2003a). The closed loop feedback simulations explore the effect of both fixed phase and variable phase feedback on the wake. While fixed phase feedback is effective in reducing drag and unsteady lift, it fails to stabilize this state once the low drag state has been reached. Variable phase feedback, however, achieves the same drag and unsteady lift reductions while being able to stabilize the flow in the low drag state. In the low drag state, the near wake is entirely steady, while the far wake exhibits vortex shedding at a reduced intensity. We achieved a drag reduction of close to 90% of the vortex-induced drag, and lowered the unsteady lift force by the same amount.

c. Validation of the Closed-Loop Controller

A low-dimensional model allows for controller development and if a more accurate non-linear model, having more modes that those used for controller development, is employed then the controller features may be analyzed as well. However, as the common saying goes "the taste of the pudding is in the eating", we need to validate the controller effectiveness in experiment. Nevertheless, a high resolution, CFD based truth simulation can provide very important insight into the complexities of feedback flow control. Both of these comprehensive approaches have been used by the USAFA team and the following are some highlights of these studies (Seidel et al, 2006, Siegel et al, 2004).

1. Siegel et al (2004) investigated the effect of feedback flow control on the wake of a circular cylinder at a Reynolds number of 100 in a water tunnel experiment. Our control approach uses a low dimensional model based on proper orthogonal decomposition (POD). The mode amplitudes are estimated in real time using Linear Stochastic Estimation (LSE) and an array of 35 sensors distributed in a stream-wise plane in the near wake. The controller applies linear proportional and differential (PD) feedback to the estimate of the first POD mode. We find the Kármán Vortex Street to be either weakened or strengthened depending on the phase shift applied by the PD controller. For all cases with a strengthening in vortex shedding, the flow becomes two-dimensional and phase locked across the entire span of the model. For all cases with a reduction in vortex shedding strength, a strong span-wise phase variation develops which ultimately leads to a loss of control even at the sensor plane location. This suggests that for reduction of vortex shedding a three-dimensional sensing and / or actuation approach is needed.

2. Siegel et al (2005b) conduct two dimensional feedback control simulations of the wake behind a D-shaped Cylinder and compare results to those obtained for the feedback controlled circular cylinder case. A POD based low dimensional model in conjunction with real time LSE is used to estimate the flow state. At laminar Reynolds numbers of up to 300, the von Kármán Vortex Street can be strengthened or weakened depending on the phase shift applied in the controller. As opposed to the circular cylinder simulations, where actuation was implemented by translating the cylinder normal to the flow, the D shaped cylinder wake is controlled using two blowing and suction slots near the base of the model. Since the D shaped cylinder features a fixed separation point, this investigation truly demonstrates that our control approach controls the wake instability and not the separation location. Results of the high resolution simulations of the feedback controlled truth model show a reduction in unsteady lift force of 40%, and a reduction in drag of 10% of the unforced flow field, using linear proportional fixed gain feedback of the first POD mode.

3. Seidel et al (2006) conduct high resolution, three-dimensional feedback controlled simulations of the wake behind a circular cylinder. In the current simulations, a three-dimensional sensor array was placed in the wake to estimate the flow state based on two dimensional POD Modes, which were applied at multiple span-wise locations. An LSE algorithm was used to map sensor readings to the temporal coefficients of the POD modes. The simulations were aimed at investigating the efficacy of three dimensional flow sensing to improve feedback control. Because the control input had only one degree of freedom (1 DOF), the mode amplitudes had to be combined into one actuator signal. Starting from an idealized, highly two-dimensional open loop case, the three-dimensional feedback controlled simulations show that, independent of the number and location of the sensor planes, control is initially successful for the whole span-wise extent. For approximately two seconds or ten vortex shedding cycles, the controller is able to significantly reduce the vortex shedding, resulting in a reduction of the drag coefficient of more than ten percent.

3.2 Experimental set-up

All tests were conducted in the USAFA Low Speed Wind Tunnel (LSWT). This tunnel has a 3 ft x 3 ft test section with a usable velocity range from 16 ft/s to 115 ft/s. A 3.5 in diameter, D, PVC cylinder spanned the entire height of the test section. Plasma actuators were placed along the span at the ±90° marks based on previous work done by List et al (2003) indicating this as the best position. The actuators consisted of two strips of copper tape, one buried beneath the dialectic barrier and one on top. Computer controlled voltage was amplified and transformers were used to significantly increase the magnitude to 11kV. The plasma formed atop the Teflon tape over the area of the buried electrode. Five layers of Teflon dielectric tape were used, as shown effective through McLaughlin et al (2006). In this case however, the Teflon tape was only used on the front side of the cylinder to make room for the sensors on the back half (Figure 3).

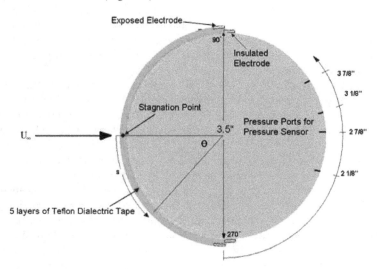

Fig. 3. Top view of cylinder set-up.

A panel was cut from the downstream side of the cylinder for sensor placement. Sixteen pressure ports consisting of four rows of four ports were placed into this panel and a Scanivalve pressure multiplexer was fixed inside the cylinder with tubing connected to each of the sixteen ports (Figure 4).

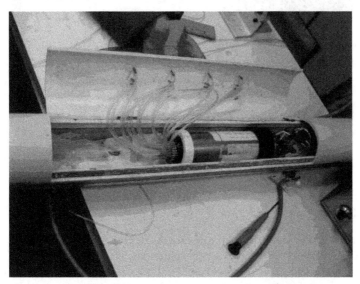

Fig. 4. Scanivalve pressure multiplexer and pressure ports in cylinder.

The location of the pressure ports was determined by doing hot film testing across the back side cylinder 1/8" behind the cylinder wall. The plasma actuators were operated at the natural shedding frequency to ensure lock-in and provide adequate flow control. Before data was collected, flow visualization was conducted to see the flow characteristics and ensure the plasma was effective in forcing the flow. Hot film anemometers were also used to validate the theoretical values for frequency downstream of the cylinder. The hot film anemometers were used to gather preliminary data very near the surface of the cylinder. The data collected was used to enable a preliminary guess in choosing pressure port locations for identifying certain flow characteristics. Figure 5 shows the tunnel set-up of the cylinder with pressure ports and the hot films positioned in the wake.

The Validyne pressure sensor was used in conjunction with a Scanivalve pressure multiplexer unit to cycle through the 16 different pressure ports. These ports were drilled into the removable rear section of the cylinder. The locations of the ports can be found in Figures 3-4. The pressure sensor has a pressure range of ±0.03 psid, an analog output of ±10Vdc, and accuracy of 0.25%. To use both the Scanivalve pressure multiplexer and Validyne sensor together, the central transducer of the Scanivalve pressure multiplexer was removed and replaced with a "dummy" plug which simply makes the Scanivalve pressure multiplexer a switching mechanism for the separate pressure ports. A period of 60 seconds was required between each pressure reading in order to ensure that the flow had "settled" after each Scanivalve pressure multiplexer switch. The remote placement of the sensor eliminated EMI issues because it was physically separated from the plasma actuators so that

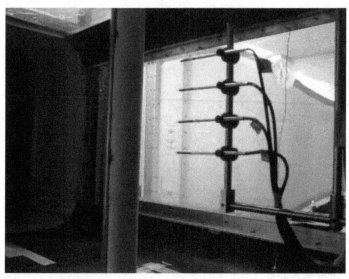

Fig. 5. Hot film anemometers and pressure ports.

id was not subject to any interference. To ensure the data acquired was not contaminated by the remote set-up, the characteristics of the plumbing were examined. For the sensor to output reliable data, the natural frequency of the plumbing system must be at least five times that of the largest frequency to be measured according to the documentation included with the sensor. The natural frequency of the system was found using the equation

$$\omega_n = \frac{c}{L\sqrt{\frac{1}{2} + \frac{Q}{aL}}} , \tag{1}$$

where ω_n is the natural frequency, c is the speed of sound (1089.2 ft/s), L is the length of tubing (2.5 ft), Q is the transducer cavity volume (2.03E-5 ft³), and a is the cross sectional area of the tubing (2.13053E-5 ft²). This yielded ω_n=73.87 Hz, which was well within the specified range since the maximum frequency measured was 9.1 Hz at Reynolds number of 20,000. This gave around 2-3% amplification of pressure waveform.

4. Results

The Validyne sensor that was connected through the Scanivalve pressure multiplexer to the pressure ports provided the surface mounted measurements required for flow state estimation. The collection of wake mounted hot film measurements and the pressure readings at each port was acquired at a sampling rate of 1 kHz. This ensured that the comparative studies could be adequately analyzed. The fundamental frequency, associated with the von Kármán vortex shedding frequency is very distinctly identified. The frequency content of the data, pertaining to the von Kármán vortex shedding frequency, from the surface mounted pressure measurements perfectly correlates with that of the hot film anemometers. For the unforced flow, it can be seen that both sensors are picking up the exact same shedding frequency of 9.1 Hz. Again, with plasma forcing within the lock-in regime, the same data are taken and using a fast Fourier transformation, the fundamental frequency is found to be very distinct. During the DBD plasma forcing, the flow's shedding frequency gets locked into the plasma actuation, which was set to a frequency of 8.8 Hz. The velocity measured by the hot films in the wake at 1.5-2.5 diameters downstream was 3-5% greater than the velocity set for the tunnel which was expected and within the range of the calculated blockage error. Since the area of the test section is reduced by the relatively large model (blockage ratio of 9.7%), the flow's velocity was increased while the resulting natural shedding frequency was also increased. Furthermore, the shedding frequency of the Re=20k flow was increased from 8.7 Hz to 9.1 Hz.

Feasible real time estimation and control of the cylinder wake may be effectively realized by reducing the model complexity of the cylinder wake using POD techniques. POD, a non-linear model reduction approach is also referred to in the literature as the Karhunen-Loeve expansion (Holmes et al, 1996). The truncated POD model will contain an adequate number of modes to enable modeling of the temporal and spatial characteristics of the large-scale coherent structures inherent in the flow. Since a pressure multiplexer is used to collect data from the 16 pressure ports, it is imperative to synchronize the time histories of the pressure measurements before any meaningful analysis of the results can be made. For this purpose, the hot film velocity measurements are used to initiate all pressure signals based on the very distinctive fundamental frequency. While this approach is inaccurate, it does provide some interesting insight into the applicability of surface mounted pressure sensors for low-order modeling of the cylinder wake at Re~20,000. In order to examine the robustness of this procedure, the POD procedure was applied to 4 snapshot sets each containing 1601, 2601, 3601 and 4601 snapshots for both plasma off and plasma on cases. The resulting Eigen-values, without and with the mean flow mode, are presented in Tables 1 and 2 respectively. It can be seen that the Eigen-value distribution is relatively insensitive to the number of snap-shots. Also, the spatial modes for plasma-off, as shown in Figure 6 (1601 snap-shots), and for plasma off (4601 snapshots), as shown in Figure 7, are fairly similar. The temporal coefficients were also found to be of a similar nature as will be discussed later. Additionally, it can be seen in Tables 1-2 that as the plasma is turned on, the intensity of the Eigen-values of modes one and two (von Kármán modes) is increased while the mean mode as well as the higher mode amplitudes are reduced.

Mode	1601 Snapshots		2601 Snapshots		3601 Snapshots		4601 Snapshots	
	Plasma Off [%]	Plasma On [%]	Plasma Off [%]	Plasma On [%]	Plasma Off [%]	Plasma On [%]	Plasma Off [%]	Plasma On [%]
1	37.99	45.97	29.04	46.01	24.00	44.88	21.25	44.27
2	17.88	31.95	20.70	31.70	21.35	32.78	17.82	33.40
3	10.58	3.38	9.82	2.93	11.51	2.68	13.12	2.45
4	6.41	3.06	7.70	2.67	7.23	2.35	7.11	2.34
5	4.12	2.50	4.76	2.28	6.09	2.09	6.65	2.09
6	3.66	2.13	4.16	2.03	4.74	1.96	5.59	1.99
7	3.31	1.92	3.83	1.92	3.67	1.92	4.64	1.85
8	2.97	1.67	3.22	1.66	3.52	1.78	3.88	1.68

Table 1. POD – Eigen-values of Surface Pressure @ Re~20K (after extraction of the mean).

Mode	1601 Snapshots		2601 Snapshots		3601 Snapshots		4601 Snapshots	
	Plasma Off [%]	Plasma On [%]	Plasma Off [%]	Plasma On [%]	Plasma Off [%]	Plasma On [%]	Plasma Off [%]	Plasma On [%]
Mean	97.38	96.74	97.25	96.73	97.36	96.74	97.37	96.76
1	0.94	1.51	0.78	1.52	0.69	1.47	0.60	1.44
2	0.51	1.06	0.56	1.05	0.48	1.09	0.41	1.11
3	0.28	0.11	0.28	0.10	0.31	0.09	0.34	0.08
4	0.17	0.10	0.23	0.09	0.21	0.08	0.20	0.08
5	0.11	0.08	0.14	0.07	0.17	0.07	0.18	0.07
6	0.10	0.07	0.12	0.07	0.13	0.07	0.16	0.06
7	0.09	0.06	0.11	0.06	0.11	0.06	0.13	0.06
8	0.08	0.05	0.09	0.05	0.10	0.06	0.10	0.05

Table 2. POD – Eigen-values of Surface Pressure @ Re~20K (after inclusion of the mean).

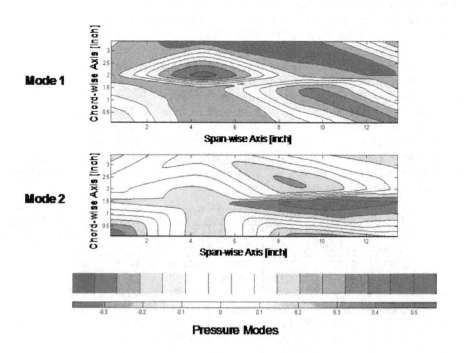

Fig. 6. First two POD Spatial Periodic Modes (1601 Snap-Shots) – Plasma Off.

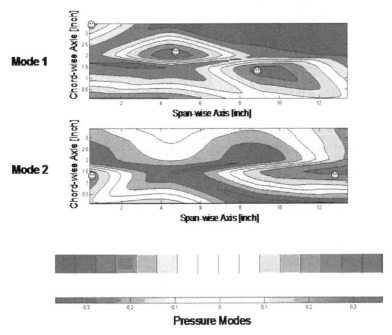

Fig. 7. First two POD Spatial Periodic Modes (4601 Snap-Shots) – Plasma On at 8.8 Hz, 11 KVolt, Position of Sensor is marked with ☺.

The spatial modes obtained from the POD procedure provide information concerning the location of areas where modal activity is at its highest. These energetic areas are the maxima and minima of the spatial modes (Cohen et al, 2006a). In this effort, 5 of the surface mounted pressure sensors, which are positioned at the energetic maxima and minima of each of the von Kármán modes, are used to provide an estimate of the POD time coefficients.

Now that the sensor configuration is determined an Artificial Neural Network Estimator (ANNE) is developed for the real-time mapping of pressure measurements onto POD time coefficients. The main features of ANNE, as described in a flow-chart and schematically in Figures 8-9, are as follows:

- **Input Layer**
- Five body mounted pressure sensor signals
- # inputs to ANN = (# past inputs per sensor) * (# sensors)*(# Time Delay) + bias
  # inputs to ANN = 4*5*4+1 = 81
- **Hidden Layer**
- 6 neurons in single hidden layer
- Activation function is based on the *tanh* function.
- A single bias input has been added
- **Output Layer**
- Three outputs, namely, the 3 POD states (A "mean flow" aperiodic mode and the two von Kármán periodic modes)
- A linear activation function.

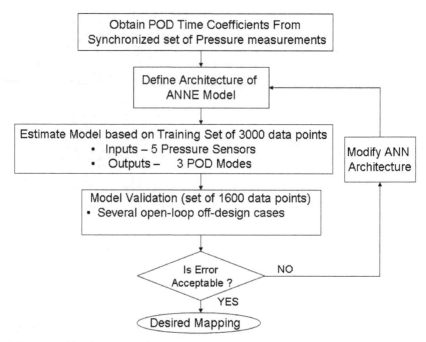

Fig. 8. Mapping of Body Mounted Pressure Measurements to POD time coefficients.

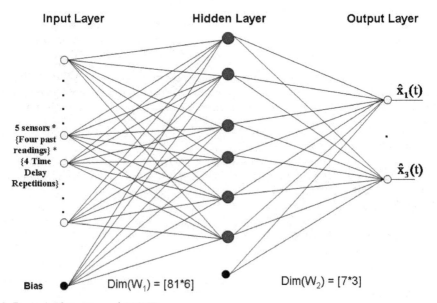

Fig. 9. Basic Architecture of ANNE.

- • **Weighting Matrices**
- The weighting matrices (W1 and W2) are initialized randomly.
- W1 between the input layer and the hidden layer is of the order of [81*6].
- W2 between the hidden layer and the output layer is of the order of [7*6].
- • **Training ANNE**
- ANNE model based on Nørgaard et al's [33] toolbox.
- Back propagation was based on the Levenberg-Marquardt algorithm.
- The training data has 3000 time steps taken at a sampling rate of 1 KHz (~26 shedding cycles)
- The training procedure converged after 250 iterations.
- • **Validating ANNE**
- Comprised of 1600 time steps taken at a sampling rate of 1 kHz (~14 shedding cycles).
- The RMS error in [%], for the 6 modes for each case was then calculated.

The estimations provided by ANNE for the 3 mode model is given in Figure 10 for the training data and in Figure 11 for the validation data. These preliminary results appear to be promising. However, one must be reminded that the main aim in this exercise is to obtain an insight for the application of the d low-dimensional suite of tools, which were primarily developed for low Reynolds laminar bluff body wakes, to higher Reynolds number turbulent wakes.

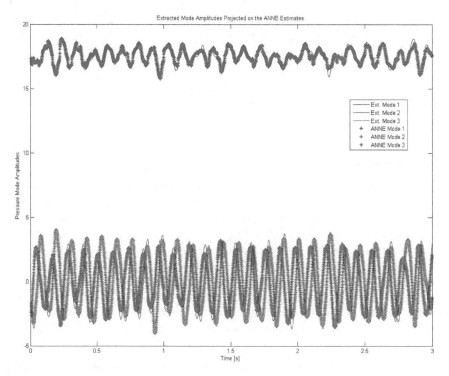

Fig. 10. Predictions based on ANNE (Training Data).

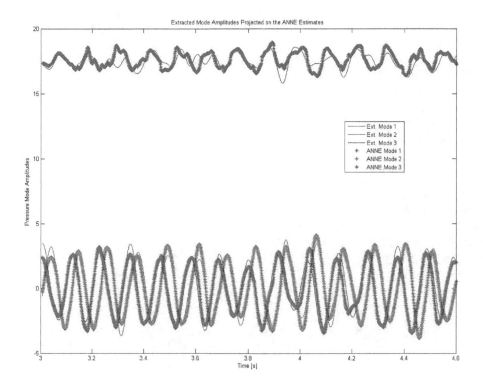

Fig. 11. Predictions based on ANNE (Validation Data).

5. Conclusions

In this chapter, we present a potentially promising approach for closed-loop flow control of a turbulent wake behind a circular cylinder at higher Reynolds numbers (Re ~ 20,000), with the ultimate goal being closed-loop flow control of the cylinder wake using DBD plasma actuators. The proposed methodology for approaching closed-loop flow control is based on the research effort at the USAF Academy over the past five years. This approach has been developed with a focus on control of laminar bluff body wakes at low Reynolds numbers (Re~50-180). The approach consists of six steps, namely: Identification of the "lock-in" region; open-loop, transient excitation using actuators; development of a low-dimensional model based on POD; sensor configuration and estimator development; development and analysis of a control law; and finally validation of the closed-loop controller.

Experimental results using plasma actuation and surface mounted pressure sensors for a circular cylinder at Reynolds number of 20,000 show that the fundamental frequency, which is paramount for feedback, is distinctly and accurately picked up by the surface mounted pressure measurements. Surface mounted pressure measurements seem to be useful for feedback of plasma forced cylinder wake at Reynolds number of 20000. Based on these experimental results, it appears that the low dimensional approach and tools developed by USAFA/DFAN for low Reynolds number (Re~100) (Sensor placement and number strategy; and ANNE estimation of the POD temporal coefficients based on surface mounted pressure sensors) feedback flow control is applicable to much higher Reynolds number (Re~20,000).

6. Acknowledgments

The authors would like to acknowledge funding by the Air Force Office of Scientific Research, LtCol Sharon Heise, Program Manager. The authors would like to thank cadets Brandon Snyder, Joshua Lewis, and Assistant Professor Christopher Seaver for preparing the experimental aspects of this effort and collecting the experimental data in the wind tunnel. The contributions of SSgt Mary Church in set-up, tear-down, and everything in between, Ken Ostasiewski and his help with the Scanivalve pressure multiplexer, and many other facets of the wind tunnel testing, and Jeff Falkenstine with the manufacturing of the model were vital to the success of the work. We would also like to thank Mr. Tim Hayden for the assistance in experimentation. The authors appreciate the discussions with Dr. Young-Sug Shin on system identification using Artificial Neural Networks. This work has been presented at an AIAA conference and published in the conference proceedings. It has no copyright protection in US since it is considered as US government work.

7. References

Adrian, R.J. (1977). On the Role of Conditional Averages in Turbulence Theory, Proceedings of the Fourth Biennial Symposium on Turbulence in Liquids, Science Press, Princeton.

Aradag, S. (2009). Unsteady Vortex Structure Behind a Three Dimensional Turbulent Cylinder Wake, Journal of Thermal Science and Technology, vol 29 (1), pp 91-98.

Aradag, S., Cohen, K, Seaver, C. and Mclaughlin, T. (2010). Integrating CFD and Experiments for Undergraduate Research, Computer Applications in Engineering Education, Vol. 18, No.4, pp 727-735.

Asghar, A., and Jumper, E.J. (2003). Phase Synchronization of Vortex Shedding from Multiple Cylinders using Plasma Actuators, AIAA Paper 2003-1028.

Ausseur, J. M., Pinier J., T., Glauser, M.N., Higuchi H., and Carlson, H. (2006). Experimental Development of a Reduced-Order Model for Flow Separation Control, AIAA Paper 2006-1251.

Bevan, B., Locke, E., Siegel, S., Mclaughlin, T.E., and Cohen, K. (2003). Flow Visualization of Three Dimensional Effects in a Cylinder Wake Flow, APS / DFD Meeting East Rutherford, NJ.

Cattafesta Iii, L.N., Williams, D.R., Rowley, C.W., and Alvi, F.S. (2003). Review of Active Control of Flow-Induced Cavity Resonance, AIAA Paper 2003-3567, 2003.

Cohen K., Siegel S., Mclaughlin T., and Gillies E. (2003). Feedback Control of a Cylinder Wake Low-Dimensional Model, AIAA Journal, 41 (8).

Cohen, K., Siegel, S., Wetlesen, D., Cameron, J., and Sick, A. (2004). Effective Sensor Placements for the Estimation of Proper Orthogonal Decomposition Mode Coefficients in von Kármán Vortex Stree, Journal of Vibration and Control, 10 (12), 1857-1880.

Cohen, K., D., Siegel, S., Luchtenburg, M., and Mclaughlin, T., And Seifert, A. (2004). Sensor Placement for Closed-loop Flow control of a 'D' Shaped Cylinder Wake, AIAA-2004-2523.

Cohen, K., Siegel S., Mclaughlin T., Gillies E., and Myatt, J. (2005). Closed-loop approaches to control of a wake flow modeled by the Ginzburg-Landau equation, Computers and Fluids, 34 (8) , 927-949.

Cohen, K., Siegel S., and Mclaughlin T. (2006a). A Heuristic Approach to Effective Sensor Placement for Modeling of a Cylinder Wake, Computers and Fluids, 35 (1), 103-120, 2006.

Cohen, K., Siegel, S., Seidel, J., and Mclaughlin, T. (2006b). System Identification of a Low Dimensional Model of a Cylinder Wake, AIAA-2006-1411.

Cohen, K., Siegel, S., Seidel, J., and Mclaughlin, T. (2006c). Neural Network Estimator for Low-Dimensional Modeling of a Cylinder Wake, AIAA-2006-3491.

Cybenko, G.V. (1989). Approximation by Superpositions of a Sigmoidal function, Mathematics of Control, Signals and Systems, 2, 303-314.

Enloe, C.L., Mclaughlin, T.E., Vandyken, R.D., Kachner, K.D., Jumper, E.J., and Corke, T.C. (2004). Mechanisms and Responses of a Single Dielectric Barrier Plasma Actuator: Plasma Morphology, AIAA Journal, 42 (3).

Gad-El-Hak, M. (1996). Modern Developments in Flow Control", Applied Mechanics Reviews, 49, 365–379.

Gerhard, J., Pastoor, M., King, R., Noack, B.R., Dillmann, A., Morzynski, M. and Tadmor, G. (2003). Model-based Control of Vortex Shedding using Low-dimensional Galerkin Models, AIAA Paper2003-4262.

Gillies, E. A. (1995). Low-Dimensional Characterization and Control of Non-Linear Wake Flows, Ph.D. Dissertation, Faculty of Engineering, Univ. Of Glasgow, Glasgow, Scotland, U.K.

Gillies, E. A. (1998). Low-dimensional Control of the Circular Cylinder Wake, Journal of Fluid Mechanics, 371, 157-178.

Glauser, M., Young, M., Higuchi, H., Tinney, C.E., and Carlson, H. (2004). POD Based Experimental Flow Control on a NACA-4412 Airfoil (Invited), AIAA Paper 2004-0574.

Haykin, S. (1999). Neural Networks – A comprehensive foundation, Second Edition, Prentice Hall, New Jersey, USA.

Holmes, P., Lumley, J.L., and Berkooz, G. (1996). Turbulence, Coherent Structures, Dynamical Systems and Symmetry, Cambridge University and Press, Cambridge.

Koopmann, G. (1967). The Vortex Wakes of Vibrating Cylinders at Low Reynolds Numbers, Journal of Fluid Mechanics, 28 (3), 501-512.

List, J., Mclaughlin, T.E., Byerley, A.R., and Van Dyken, R.D. (2003). Using Plasma to Control Separation on a Linear Cascade Turbine Blade, AIAA Paper 2003-1026.

Mclaughlin, T., Felker, B., Avery, J. and Enloe, C. (2006). Further Experiments in Cylinder Wake Modification With Dielectric Barrier Discharge Forcing, AIAA-2006-1409.

Munska, M. And Mclaughlin, T.E. (2005) Circular Cylinder Flow Control using Plasma Actuators, AIAA Paper 2005-0141.

Murray, N.E., and Ukeiley, L.S. (2002). Estimating the Shear Layer Velocity Field Above an Open Cavity from Surface Pressure Measurements, AIAA Paper 2002-2866.

Nelles, O. (2001). Nonlinear System Identification, Springer-Verlag, Berlin, Germany.

Noack, B.R., Afanasiev, K., Morzynski, M. And Thiele, F. (2003). A hierarchy of low-dimensional models for the transient and post-transient cylinder wake, J. Fluid Mechanics, 497, 335–363.

Noack, B.R., Tadmor, G., and Morzynski, M. (2004). Low-dimensional models for feedback flow control. Part I: Empirical Galerkin models, AIAA-2004-2408.

Nørgaard, M., Ravn, O., Poulsen, N. K., and Hansen, L. K. (2000). Neural Networks for Modeling and Control of Dynamic Systems, Springer-Verlag, London, UK.

Rapoport, D., Fono, I., Cohen, K., and Seifert, A. (2003). Closed-loop Vectoring Control of a Turbulent Jet Using Periodic Excitation, Journal of Propulsion and Power, 19 (4), 646-654.

Samimy, M., Debiasi, M., Caraballo, E., E.C., Ozbay, H., Efe, M.O., Yuan, X., Debonis, J., and Myatt, J.H. (2003). Development of Closed-loop Control for Cavity Flows, AIAA Paper 2003-425.

Seidel, J., Siegel, S., Cohen, K., and Mclaughlin, T. (2006a). Simulations of Flow Control of the Wake behind an Axisymmetric Bluff Body, AIAA-2006-3490.

Siegel S., Cohen, K. and Mclaughlin T. (2003a). Feedback Control of a Circular Cylinder Wake in Experiment and Simulation, AIAA Paper 2003-3569.

Siegel, S., Cohen, K., and Mclaughlin, T. (2004). Experimental Variable Gain Feedback Control of a Circular Cylinder Wake, AIAA-2004-2611.

Siegel S., Cohen K., Seidel, J., and Mclaughlin T. (2005a). Short Time Proper Orthogonal Decomposition for State Estimation of Transient Flow Fields, AIAA-2005-0296.

Siegel S., Cohen K., Seidel, J., and Mclaughlin T. (2005b). Two Dimensional Simulations of a Feedback Controlled D-Cylinder Wake, AIAA-2005-5019.

Sirovich, L. (1987). Turbulence and the Dynamics of Coherent Structures Part I: Coherent Structures, Quarterly of Applied Mathematics, 45 (3), 561-571.

Stalnov, O., Palei, V., Fono, I., Cohen, K., and Seifert, A. (2005). Experimental Validation of Sensor Placement for Control of a "D" Shaped Cylinder Wake, AIAA 2005-5260.

Taylor, J.A. and Glauser, M.N. (2004). Towards Practical Flow Sensing and Control via POD and LSE Based Low-Dimensional Tools, Journal of Fluids Engineering, 126, 337-345.

6

Modeling the Wake Behind Bluff Bodies for Flow Control at Laminar and Turbulent Reynolds Numbers Using Artificial Neural Networks

Selin Aradag and Akin Paksoy

TOBB University of Economics and Technology,
Turkey

1. Introduction

The phenomenon of vortex shedding behind bluff bodies has been a subject of extensive research. Many flows of engineering interest produce this phenomenon and the associated periodic lift and drag response (Cohen et al., 2005). External flow over bluff bodies is an important research area because of its wide range of engineering applications. Although, the geometry of a bluff body can be simple, the flow behind it is chaotic and time-dependent after a certain value of Reynolds number. Forces acting on the body such as drag and lift also vary in time, and cause periodic loading on it. These forces originate from momentum transfer from fluid to the body, where their magnitudes are strongly related to the shape of the body and properties of the flow.

Flow over a circular cylinder is a benchmark problem in literature. It arises in diverse engineering applications such as hydrodynamic loading on marine pipelines, risers, offshore platform support legs, chemical mixing, lift enhancement etc. (Gillies, 1998; Ong et al., 2009). It is experimentally investigated by Norberg (1987) that when the Reynolds number of flow over a circular cylinder exceeds 48, vortices separate from the cylinder surface, and start to move downstream, where steady-state behavior of the flow turns into a time-dependent state. These periodically moving vortices at the downstream form self-excited oscillations called the von Kármán vortex street (Gillies, 1998) as shown in Fig. 1.

Fig. 1. The von Kármán vortex street observed in the wake region of a two-dimensional circular cylinder

Separation from the surface of the cylinder can be either laminar or turbulent according to the regime of the flow in the boundary layer. It is shown by Wissink and Rodi (2008) that flow with a Reynolds number between 1000 and 20000 is called subcritical, and in this range, boundary layer on the cylinder is entirely laminar and transition from laminar to turbulent flow happens somewhere at the downstream. Although vortex street is fully turbulent after Re≈20000, laminar separation sustains up to a Reynolds number of 100000 (Travin, 1999). Several experimental and computational studies in literature examine the flow over a circular cylinder at subcritical Reynolds numbers (Anderson, 1991; Aradag, 2009; Aradag et al., 2009; Lim and Lee, 2002).

The control of the vortex shedding observed in the wake region of a bluff body is extremely important in engineering applications in order to improve aerodynamic characteristics and performance of the bluff body. To do this, it is substantially important to predict the flow structures and their characteristics observed in the wake region (Aradag, 2009).

In many of the engineering applications involving fluids, Computational Fluid Dynamics (CFD) plays a crucial role as a major tool to analyze flow structures and their characteristics (Gracia, 2010). However, it lacks the functionality of being practical and quick for real-time complex fluid mechanics applications, and such limitations cause difficulties especially in the development of flow control strategies (Fitzpatrick et al., 2005). In order to observe the flow structures and their characteristics in real-time systems in detail, a more practical procedure is needed.

The Proper Orthogonal Decomposition (POD) is a reduced order modeling technique used to analyze experimental and computational data by identifying the most energetic modes and relative mode amplitudes in a sequence of snapshots from a time-dependent system (Cao et al., 2006). It has been used in numerous applications to introduce low-dimensional descriptions of system dynamics by extracting dominant features and trends (Lumley, 1967). The POD technique was originally developed in the context of pattern recognition, and it has been used successfully as a method for determining low-dimensional descriptions for human face, structural vibrations, damage detection and turbulent fluid flows (Chatterjee, 2000). In addition, the method has also been used for many industrial and natural applications, such as supersonic jet modeling, thermal processing of foods, investigation of the dynamic wind pressures acting on buildings, weather forecasting and operational oceanography (Cao et al., 2006). There are several studies in literature that utilize the POD technique in fluid mechanics applications as a reduced order modeling tool (Connell & Kulasiri, 2005; Lieu et al., 2006; O'Donnell & Helenbrook, 2007; Sen et al., 2007; Unal & Rockwell, 2002).

In POD technique, originally correlated data is linearly combined to form principal components that are uncorrelated and ordered according to the portion of the total variance in the considered data (Samarasinghe, 2006). This type of dimensionality reduction offers linear combinations of orthogonal functions to represent a process or a system. Thus, the order of the original high-dimensional data is reduced by compressing the essential information to the uncorrelated principal components associated with modes and relative mode amplitudes to provide a model of the data instead of using the original correlated inputs (Newman, 1996b).

The selected principal components, and hence modes and relative mode amplitudes, can be used as an alternative to the original data ensemble at the input section to a neural network. Since the number of inputs to the model is substantially reduced, the formed network structure will have less complexity and prevent overfitting while representing the original inputs appropriately (Samarasinghe, 2006).

Artificial Neural Networks (ANN's) refer to computing systems the main idea of which is inspired from the analogy of information processing in biological nervous systems. A neural network structure transforms a set of input variables into a set of output variables via mathematical and statistical approaches (Bishop, 1994). By using ANN's, it is possible to obtain a solution for complex problems that do not have an analytical solution via application of conventional approaches.

Currently, neural networks are used for the solution of problems in system identification, such as pattern recognition, data analysis, and control. Apart from these, ANN's have also been applied in diverse fields such as insurance, medicine, economic predictions, speech recognition, image processing, and heat transfer and fluid mechanics applications (Nørgaard et al., 2000). For example, in a study performed by Xie et al. (2009) ANN's are used to evaluate friction factors in shell and tube heat exchangers by making use of experimental and computational Nusselt numbers obtained at laminar and turbulent regimes, where Reynolds number changes within the range of 100 and 10000. The authors related 12 different input sets including geometric parameters, such as number of tubes, arrangement of tubes and fin structures, with Nusselt numbers and friction factors to train the feed-forward backpropagation ANN structure and to predict friction factors for similar geometries. They stated the success of the practical use, easiness and importance of ANN's by achieving only 4% difference between the original data and predicted values.

In another study accomplished by Zhang et al. (1996), ANN's are used to estimate flow characteristics by making use of previously obtained flow dynamics characteristics. The authors observed two-dimensional (2D) von Kármán vortex structures in an elongated rectangular cross-sectional area of a static prism where Reynolds number varies within 250 and 800. They used von Kármán structural phases observed at certain Reynolds numbers as previous cases for prediction of vortex formation phases for new Reynolds numbers. The developed model shows that ANN's provide significant advantages for dealing with flow problems that involve certain amount of complexity to observe flow characteristics without requiring further CFD analyses.

Several notable features of ANN's include relatively high processing speeds, ability of learning the solution of a problem from a set of examples, dealing with imprecise, noisy, and highly complex nonlinear data, and parallel processing (Khataee, 2010). These unique properties make ANN's eligible for prediction of the flow structures and their characteristics in real-time systems for development of flow control strategies.

2. Aims and concerns

The aim of this research is to represent the flow behind a two-dimensional (2D) circular cylinder at laminar and turbulent Reynolds numbers (Re), where Re=100 for the laminar and Re=20000 for the turbulent regime analyses, with the help of Artificial Neural Networks (ANN's) in order to be able to control the vortex shedding formed in the wake region. The flow analyses over the 2D circular cylinder are performed by Computational Fluid Dynamics (CFD), and the results are validated with the experimental results given in literature. In order to observe laminar and turbulent flow structures and their effects in the wake region for control purposes with ANN's, orders of the original CFD data ensembles containing the x-direction velocities at each nodal point of the grids are reduced by application of the Proper Orthogonal Decomposition (POD) technique.

For laminar flow POD analyses, the classical "Snapshot Method" developed by Sirovich (1987) is used; however, for turbulent flows this method causes certain drawbacks, such as lacking the ability of separating flow structures according to their scales during configuration of the modes and relative mode amplitudes. Since it is inevitable to use the POD technique to obtain a low-dimensional description of the original data ensembles for further ANN applications, the classical "Snapshot Method" is combined with the Fast Fourier Transform (FFT) filtering procedure for turbulent flow POD analyses as suggested by Aradag et al (2010). The combined FFT-POD technique is performed to the turbulent CFD data ensembles to eliminate the undesired effects of small scale turbulent structures in the wake region, and to observe flow characteristics in more detail by separating spatial (modes) and temporal (mode amplitudes) structures. (Apacoglu, 2011b)

For real-time flow control applications, it is important to predict the flow based on surface sensors placed at a few discrete points and to relate sensor data as an input to the input section of the neural network structure (Apacoglu et al., 2011a). For this purpose, a sensor placement study is also performed to obtain optimum sensor locations on the 2D circular cylinder surface by using a one-dimensional (1D) classical POD analysis based on surface pressure data of the CFD results. (Apacoglu et al, 2011a)

ANN's are used to predict the temporal structures (mode amplitudes) obtained from the POD and the FFT-POD analyses respectively for laminar and turbulent flow cases by using only the sensor data from several locations on the 2D circular cylinder surface. The training and validation data used for the neural network structure are from several computational cases. Consequently, the defined ANN approach helps to predict what is happening in the flow without requiring further CFD simulations, which are very expensive and impossible in real-time flow control applications. This chapter summarizes the ANN based modeling of the flow structures behind a 2D circular cylinder based on the CFD and POD results given by Apacoglu et al (2011a) for laminar flow and Apacoglu et al (2011b) for turbulent flow. The results obtained in these two articles are used as inputs for training the neural nets in this work.

3. Research methods

3.1 Computational Fluid Dynamics (CFD) methodology

The details on the boundary conditions, grid refinement study and computations are provided in Apacoglu et al. (2011a, 2011b). Operating conditions for the simulations are given in Table 1. The drag coefficient (C_D), Strouhal number (St), pressure coefficient distribution around the cylinder and the velocity profiles at the wake are validated using the experimental results of Lim and Lee (2002), and Aradag (2009).

	Laminar Flow	Turbulent Flow	
Parameter	Value	Value	Unit
Reynolds number	100	20000	-
Density	5.25×10^{-5}	0.01056	kg/m^3
Free-Stream Velocity	34	34	m/s
Viscosity	1.78×10^{-5}	1.795×10^{-5}	kg/ms
Pressure	4.337	872.36	Pa

Table 1. Operating conditions for the flow simulations (Apacoglu et al, 2011a, Apacoglu et al, 2011b)

The 2D circular cylinder is designed to comprise four slots on its surface to force the flow by air blowing as shown in Fig. 2. (Apacoglu et al, 2011a)

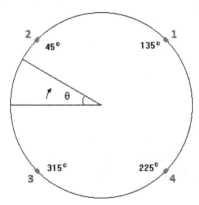

Fig. 2. Position of the slots located on the circumference of the cylinder (Apacoglu et al, 2011a)

Slots are either closed or opened in different combinations at different blowing velocities as outlined below:

- Blowing from all slots, u=0.1U
- Blowing from all slots, u=0.5U
- Blowing from slot numbered 1, u=0.1U
- Blowing from slots numbered 1 and 2, u=0.1U
- Blowing from slots numbered 1 and 4, u=0.1U
- Blowing from slots numbered 2 and 3, u=0.1U
- Blowing from slots numbered 1 and 4, u=0.5U (only at turbulent flow regime)

where u represents blowing velocity and the free stream velocity U=34 m.s^{-1}. In order to obtain data ensemble required for the POD and FFT-POD applications, and ANN estimations, velocity values obtained for the x-direction at the wake region are recorded in each time step for 10 periods of the flow time. (Apacoglu et al, 2011a, Apacoglu et al, 2011b)

3.2 Proper Orthogonal Decomposition (POD) and filtering methodology

The POD technique is applied to CFD data ensembles containing x-direction velocity magnitudes observed in the wake region of the 2D circular cylinder in either laminar or turbulent regimes as a post processing to tool to reduce the order of the data and prepare them for further ANN applications. (Apacoglu et al, 2011a, Apacoglu et al, 2011b).

The originally correlated CFD data ensembles are processed to form principal components in space (modes) and time (mode amplitudes). Detailed mathematical and theoretical information is given in Newman (1996a and 1996b), Holmes et al. (1996), Ly and Tran (2001), Sanghi and Hasan (2011) and Smith et al. (2005).

3.3 Sensor placement methodology

Since the ultimate aim of this study is to control the von Kármán vortex street observed in the wake region and to estimate the state of the flow with the help of ANN's, a one-dimensional (1D) classical snapshot-based POD analysis is carried out on static pressure data obtained directly from the cylinder surface to identify optimal sensor locations. Static pressure data coming from the sensors are essential in training and simulation processes of the neural networks for enabling them to make real-time estimations (Apacoglu et al., 2011).

In practical engineering applications, it is not feasible to place and fix sensors in the wake region and to obtain accurate enough data. On the contrary, body-surface mounted sensors are simple, relatively inexpensive and provide reliable data for further analyses (Seidel et al., 2007).

Sensors identified at optimal locations provide static pressure data, which has the highest activity in terms of pressure on the cylinder surface. This case is demonstrated by an example in a study performed by De Noyer (1999). Since one prominent feature of the POD technique is to extract dominant characteristics of the data, utilization of it to static pressure data coming from the cylinder surface enable optimal locations, which are dominant in terms of pressure for sensor placement.

For laminar and turbulent flow sensor placement POD analyses, CFD data providing static pressure signals on the cylinder surface at 360 locations with one-degree increments are used. In the context of sensor placement studies, uncontrolled flow test case (all slots closed) and the most effective controlled flow test case (all slots open with 0.5U air blowing) are considered for both laminar and turbulent regimes. The details of sensor placement for the laminar case is given by Apacoglu et al (2011a).

For 1D POD analyses, laminar flow test cases (uncontrolled and 0.5U air blowing controlled) include 1800 snapshots, whereas uncontrolled and 0.5U blowing controlled turbulent flow test cases include 1337 and 1320 snapshots respectively. Table 2 shows energy contents of the most energetic four and six pressure-based POD modes turbulent flow. Energy content of each mode represents the level of dominant pressure characteristic trends monitored by that mode.

Mode Number	Energy Contents (%)		Mode Number	Energy Contents (%)	
	Uncontrolled Flow	Controlled Flow		Uncontrolled Flow	Controlled Flow
1	92.14	91.50	4	0.46	0.09
2	4.50	5.43	5	0.29	0.02
3	2.53	0.23	6	0.05	0.01
Total (3 Modes)	99.17	97.16	Total (6 Modes)	99.97	97.28

Table 2. Energy contents of the most energetic four pressure-based POD modes for uncontrolled and controlled turbulent flow test cases

Sensor locations correlated to the energetic surface pressure maxima and minima of pressure-based POD modes are given in Table 3 for turbulent flow test case. The locations of the sensors in Table 3 are referenced in terms of the circumferential angle measured from front stagnation point along the clockwise direction.

For practical applications, it is desirable to reduce the amount of sensors required for the real-time estimation of the systems (Seidel et al., 2007). Since the contributions of the most energetic two and three pressure-based POD modes to the total energy content is greater

than others respectively for laminar and turbulent flow test cases, it is concluded that taking
into account only those modes for identification of optimal sensor locations is enough.

Mode Number	Sensor Locations (degrees of angle)			
	Uncontrolled Flow		Controlled Flow	
	Θ_1	Θ_2	Θ_1	Θ_2
1	274	87	274	87
2	224	134	52	307
3	213	184	184	134
4	161	197	177	202
5	161	184	196	161
6	188	202	171	188

Table 3. Sensor locations corresponding to the minimum (Θ_1) and maximum (Θ_2) values of
the pressure-based POD modes for uncontrolled and controlled turbulent flow test cases

In addition, when Table 3 is examined in detail, it can be observed that sensor locations
corresponding to the most energetic pressure-based POD mode are not affected from air
blowing with 0.5U from the slots. Optimal sensor locations for turbulent flow test cases are
shown in Fig. 3. The sensors corresponding to the first modes (the most energetic ones)
target the periodic modes associated with the von Kármán shedding frequency, whereas the
sensors related with other modes target the non-periodic POD modes (Seidel et al., 2007).

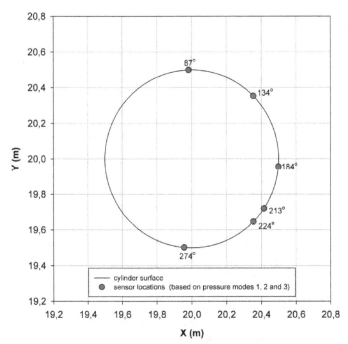

Fig. 3. Optimal sensor locations on the circumference of the cylinder for turbulent flow test cases

3.4 Artificial Neural Network (ANN) methodology

An Artificial Neural Network (ANN) is an interconnected assembly of simple processing elements, the functionality of which is loosely based on the biological neuron. The processing ability of the ANN is stored in the interunit connection strengths, or weights, obtained by a process of adaptation to, or learning from, a set of training patterns (Gurney, 1997).

In ANN, a neuron is a processing element that takes number of inputs, weights them, sums them up, and uses the result as the argument for a singular valued function, which is called the activation function (Nørgaard et al., 2000). Among a variety of network structures the most common one is the multilayer perceptron (MLP) network or also referred as the feedforward network that consists of two or more layers as shown in Fig. 4.

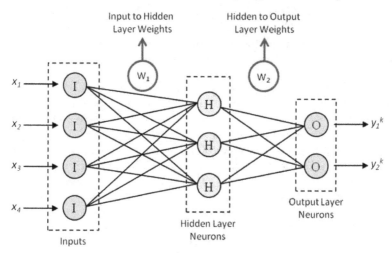

Fig. 4. Schematic representation of the basic structure of an MLP network containing one hidden layer

In Fig. 4, the first layer is known as the hidden layer since it is in some sense hidden between the external inputs (x_1 to x_4) and the output layer, which produces the output of the network (y_{1k} and y_{2k}). W_1 and W_2 are the matrices represent the weight values respectively connecting inputs to hidden layer neurons and correspondingly to output layer neurons. In order to determine the weight values included in W_1 and W_2, there has to be a set of examples of the outputs that are related to the inputs. The determination process of weights from the prior examples is known to be training or learning (Nørgaard et al., 2000; Samarasinghe, 2006).

The MLP neural network structure presents great harmony for discrete-time modeling of nonlinear dynamic systems. Especially turbulent flow systems can be counted as a major example for nonlinear dynamic systems, where the inputs to the network are related to the outputs in a highly nonlinear fashion.

Under some conditions, success of the MLP network structure may be affected negatively from one or more temporal behaviors that the system introduces during identification of the nonlinear relationships and prediction of the time series results by itself. In order to prevent such undesirable drawbacks and to provide accurate enough predictions, the MLP network structure is supplied with a short-term memory dynamics approach. This kind of neural network structures are called as Spatio-Temporal Time-Lagged Multi Layer Perceptron networks, and they can be thought of as a nonlinear extension of an auto regressive model with exogenous input variables (Samarasinghe, 2006).

In this study, the ANN estimation method of choice including application of the MLP network structure based on a nonlinear system identification in collaboration with Auto-Regressive eXternal input (ARX) model structure approach described by Norgaard et al. (2000) is used. This model includes nonlinear optimization techniques based on the Levenberg-Marquardt back propagation method. The Levenberg-Marquardt method minimizes the difference between the extracted POD mode amplitudes and the ANN estimations, while adjusting the weights of the model.

The Levenberg-Marquardt method is a hybrid algorithm that combines the advantages of the steepest descent and Gauss-Newton methods to produce a more efficient method than either of these two methods does individually. Due to its inherent property related with the conditioning parameter, the Levenberg-Marquardt method adjusts this parameter automatically in every iteration to reduce the error gradually (Samarasinghe, 2006).

The importance of the ARX engaged ANN dynamic network model structure is its strong stability capability even if the dynamic system under investigation is unstable. The stability task is at the highest level of importance when dealing with nonlinear systems of partial differential equations, such as the Navier-Stokes equations (Nørgaard et al., 2000; Siegel et al., 2008).

In this study, pressure data obtained from surface sensors and previously obtained POD or FFT-POD mode amplitudes are used as inputs to the neural network structure. At the end of ANN studies, it is needed to estimate mode amplitudes that are the same as the mode

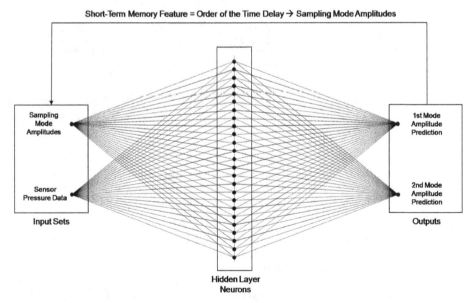

Fig. 5. Schematic representation of the neural network structure formed for analyses

amplitudes obtained from the POD analysis of the CFD results, but without using further CFD simulations. The system consists of multi inputs (sensor pressure data and sampling mode amplitudes coming from short-term memory), and requires multi outputs (each estimated mode amplitude will be an output) as shown in Fig. 5.

Further information about the basics of ANN's, different network structures and applications are given in Haykin (1994), Mehrotra et al. (2000), Samarasinghe (2006), Nørgaard et al. (2000) and Gurney (1997).

4. Results

4.1 Proper Orthogonal Decomposition (POD) and filtering results

The details of the proper orthogonal decomposition analysis are provided in Apacoglu et al (2011a) and Apacoglu et al (2011b)

Figure 6 presents relative FFT-POD mode amplitudes with respect to snapshot number for the uncontrolled (all slots closed) and the most effective controlled flow test case (all slots open with 0.5U air blowing).

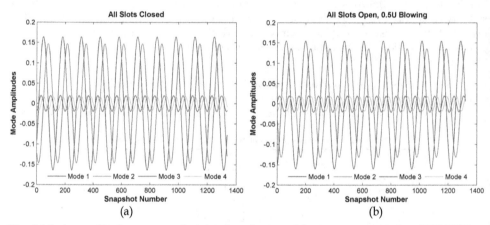

Fig. 6. Mode amplitudes vs. snapshot number change of the most energetic four FFT-POD modes for a) the uncontrolled (all slots closed) and b) the most effective controlled flow test case (all slots open with 0.5U air blowing)

In Fig. 6, since the most energetic parts of the flow characteristics are related with the modes 1 and 2 in both cases, their amplitudes are greater than modes 3 and 4. All the relative mode amplitudes show periodic behavior, which is directly associated with the existence of the von Kármán vortex street in the wake region of the 2D circular cylinder.

Another important result is that formations of the sinus curves in Fig. 6 are different from each other. For a fixed snapshot number, maximum and minimum values of the mode amplitudes show distinction. This leads to a conclusion that the vortex formation is lagged due to air blowing. By changing air blowing velocity from the slots located on the surface of the cylinder, it is possible that one can bear order of the vortex lagging to desired levels efficiently. More information on mode amplitudes and the results for laminar flow test cases may be found in Apacoglu et al. (2011a) and Paksoy et al. (2010).

4.2 Artificial Neural Network (ANN) results

There are two different Spatio-Temporal Time-Lagged Multi Layer Perceptron networks are formed to be used for laminar and turbulent flow test cases separately. Both network structures are designed to estimate the most energetic two mode amplitudes for different test cases by making use of the specified data sets employed in the training processes.

The generated ANN structures have identical properties. For example, they consist two layers (one hidden and one output) apart from the inputs sections as shown in Fig. 4. The activation neuron function is based on the nonlinear *tanh* function for both networks, and a single bias input has been added to the output from the hidden layer. The output layer has a linear activation function, and it consists of two outputs, namely the most energetic two mode amplitudes.

Both of the designed networks use a supervised learning (training) process with an adequate set of data that constitutes to approximately first half of the 10 shedding cycles. The training process uses cylinder surface pressure data obtained from the six sensors being as one set of the inputs and the sampling mode amplitudes being as the other set of the inputs, which are directly related with the order of the time delay parameter and short-term memory feature of the networks. Thus, the input sections to the networks comprise two different sets of data. After the training process, a validation step is employed by estimating the remaining data (corresponding to last five shedding cycles) to check accuracy and prediction capability of each network.

The complexity and size of the both networks can be adjusted by varying time delay and hidden layer neuron number parameters. The time delay value is directly associated with the order of the short-term memory feature. It qualifies the number of mode amplitudes that

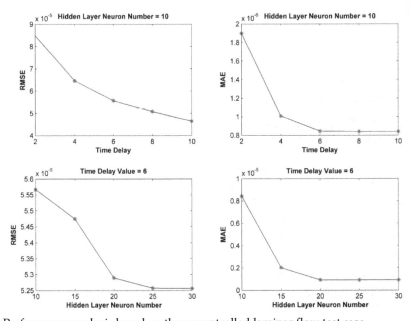

Fig. 7. Performance analysis based on the uncontrolled laminar flow test case

need to be estimated and provided to the inputs section as data observed at the previous sampling instant in addition to the sensor pressure data, which is provided externally to the networks. The hidden layer neuron number is another important parameter that influences prediction accuracy of the estimated mode amplitudes (Paksoy & Aradag, 2011). In order to acquire feasible values for the time delay and the hidden layer neuron number parameters, performances of the networks are monitorized by considering the root mean square errors (RMSE) and mean absolute errors (MAE) between the network prediction results and the target values for a couple of trials. Figures 7 and 8 present network performance analyses based on the uncontrolled flow test cases respectively for laminar and turbulent regimes.

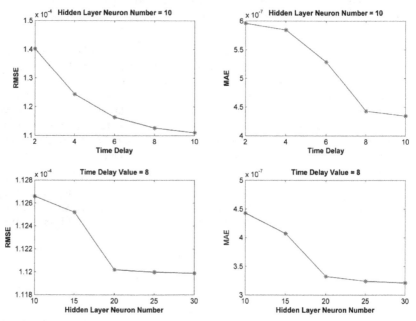

Fig. 8. Performance analysis based on the uncontrolled turbulent flow test case

As shown in Fig. 7 and Fig. 8, an increase in time delay value positively affects accuracy of the results, and relatively decreases the order of the error signals. For larger time delay values, there is more data available for the network to train itself by interconnecting the input sets via setting up larger weighing matrices, and hence weights, by making use of more known data coming from the past. However, this increases complexity of the network structure, and the time required for analyses rises.

According to results observed in Fig. 7 and Fig. 8, values of the time delay and the hidden layer neuron number are respectively specified as 6 and 25 for the laminar flow network structure, 8 and 25 for the turbulent flow network structure.

Taking into consideration of POD (applied for laminar flow test cases) and FFT-POD (applied for turbulent flow test cases) results, it is revealed that more than 90% of the total energy content can be represented by using only the two most energetic modes (1 and 2), where most of the flow structures and their characteristics are retained. For control purposes, estimations of the mode amplitudes related with those two most energetic modes plays a crucial role in effective observation of the effects flow structures and their characteristics in the flow field without requiring further CFD simulations.

ANN estimations of the mode amplitudes and their comparison with the original data for modes 1 and 2 are shown in Fig. 9 and Fig. 10 so as to observe the convenience of the validation processes for the designed network structures. Uncontrolled flow test cases of both laminar and turbulent flow analyses are selected to be used in the validation process. It can be observed from Fig. 9 and Fig. 10 that the resulting ANN estimations for the validation step show adequate coherency including only minor errors.

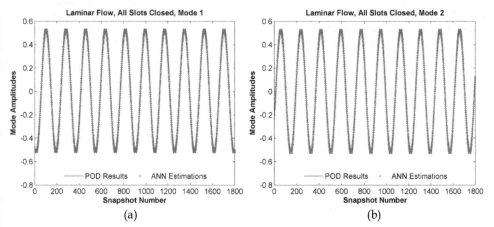

Fig. 9. Validation process ANN estimations and their comparison with the POD results of the uncontrolled laminar flow test case a) for relative mode amplitude 1 and b) for relative mode amplitude 2 with time delay 6 and hidden layer neuron number 25

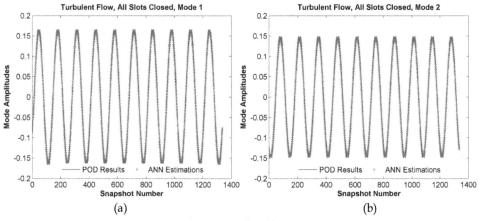

Fig. 10. Validation process ANN estimations and their comparison with the POD results of the uncontrolled turbulent flow test case a) for relative mode amplitude 1 and b) for relative mode amplitude 2 with time delay 8 and hidden layer neuron number 25

In order to see the modeled network structures in action with the specified design parameters, the networks are adjusted to estimate mode amplitudes for the controlled flow test cases in both laminar and turbulent flow analyses.

For the new estimation cases, different from the validation processes, network structures are trained with the sensor pressure data and sampling mode amplitudes belonging to the all slots open with 0.5U air blowing controlled flow test case for further laminar and turbulent ANN analyses. After training the networks with the specified controlled flow test cases, predictions are done for other controlled flow test cases by just feeding the sensor pressure data regarding to each test case as external input sets.

Figures 11, 12 and 13 show ANN predictions and original mode amplitudes (obtained in the course of POD analyses for laminar flow test cases and FFT-POD analyses for turbulent flow test cases) for a couple of selected sample test cases. Among others, the selected ones exhibit the next most effective control approach with air blowing after the all slots open with 0.5U air blowing controlled flow test case.

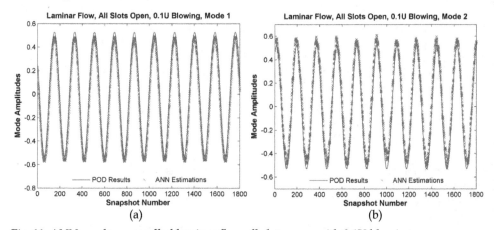

Fig. 11. ANN results, controlled laminar flow all slots open with 0.1U blowing

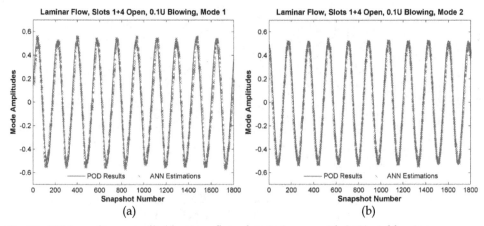

Fig. 12. ANN results, controlled laminar flow slots 1+4 open with 0.1U air blowing

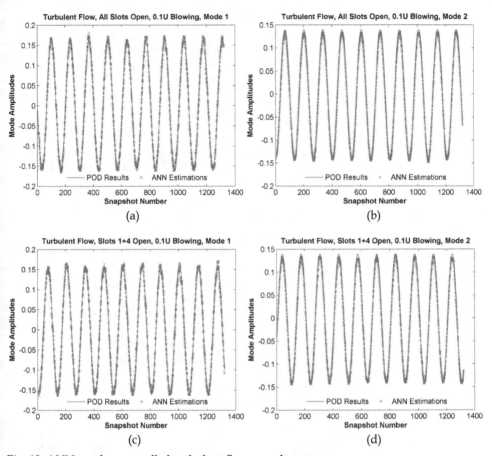

Fig. 13. ANN results, controlled turbulent flow sample test cases

Figures 11, 12 and 13 show that the results obtained from ANN estimations are in good agreement with the results obtained from the POD and the FFT-POD applications. Low-levels of acceptable ANN estimation errors are especially clustered at certain snapshot values corresponding to the lower and upper end tips of the periodic curves.

5. Conclusions

Within the scope of this study, the flow behind a 2D circular cylinder at laminar (Re=100) and turbulent (Re=20000) Reynolds numbers (Re) with the help of Artificial Neural Networks (ANN's) in order to be able to control the vortex shedding formed in the wake region.

For real-time flow control applications, in order to estimate the state of the flow, it is essential to predict the mode amplitudes regarding to the most energetic two modes. ANN's are used to predict mode amplitudes by using only the sensor data from several locations on the 2D circular cylinder surface. By implementation of the Spatio-Temporal Time-Lagged

Multi Layer Perceptron network structures, robust and real-time estimators of mode amplitudes necessary for observation of the effects of flow structures and their characteristics in the flow field are evaluated effectively without requiring further CFD simulations.

6. Acknowledgement

This research is supported by The Scientific and Technological Research Council of Turkey (TUBITAK) under grant 108M549 and Turkish Academy of Sciences Distinguished Young Scientists Awards Programme (TUBA-GEBIP). Major parts of this study were published as a conference proceeding in the 49th AIAA Aerospace Sciences Meeting held in Orlando, Florida in January, 2011.

7. References

Anderson, J.D. (1991). *Fundamentals of Aerodynamics* (2nd Edition), McGraw Hill, New York

Apacoglu, B.; Paksoy, A. & Aradag, S. (2011a). CFD Analysis and Reduced Order Modeling of Uncontrolled and Controlled Laminar Flow over a Circular Cylinder. *Engineering Applications of Computational Fluid Mechanics*, Vol. 5, No. 1, pp. 67-82

Apacoglu, B.; Paksoy, A. & Aradag, S. (2011b), Effects of Air Blowing on Turbulent Flow over a Circular Cylinder, Journal of Thermal Science and Technology, *Journal of Thermal Science and Technology*, in press

Aradag, S. (2009). Unsteady Turbulent Vortex Structure Downstream of a Three Dimensional Cylinder. *Journal of Thermal Science and Technology*, Vol. 29, No. 1, pp. 91-98

Aradag, S.; Cohen, K.; Seaver, C. & McLaughlin, T. (2009). Integrating CFD and Experiments for Undergraduate Research. *Computer Applications in Engineering Education*, doi: 10.1002/cae.20278

Aradag, S.; Siegel, S.; Seidel, J.; Cohen, K. & McLaughlin, T. (2010). Filtered POD-Based Low-Dimensional Modeling of the Three-Dimensional Turbulent Flow Behind a Circular Cylinder. *International Journal for Numerical Methods in Fluids*, doi: 10.1002/fld.2238

Bishop, C.M. (1994). Neural Networks and Their Applications, Review Article. *Review of Scientific Instruments*, Vol. 65, pp. 1803-1832

Cao, Y.; Zhu, J.; Luo, Z. & Navon, I. (2006). Reduced Order Modeling of the Upper Tropical Pacific Ocean Model Using Proper Orthogonal Decomposition. *Computer and Mathematics with Applications*, Vol. 52, No. 8-9, pp. 1373-1386

Chatterjee, A. (2000). *An Introduction to the Proper Orthogonal Decomposition, Computational Science Section Tutorial*. Department of Engineering Science and Mechanics, Penn State University, Pennsylvania

Cohen, K.; Siegel, S.; McLaughlin, T.; Gillies, E. & Myatt, J. (2005). Closed-Loop Approaches to Control of a Wake Flow Modeled by Ginzburg-Landau Equation. *Computers and Fluids*, Vol. 34, pp. 927-949

Connell, R. & Kulasiri, D. (2005). Modeling Velocity Structures in Turbulent Floods Using Proper Orthogonal Decomposition, *Proceedings of International Congress on Modeling and Simulation*, Melbourne, Australia, December 2005

De Noyer, B. (1999). *Tail Buffet Alleviation of High Performance Twin Tail Aircraft using Offset Piezoceramic Stack Actuators and Acceleration Feedback Control*, Ph.D. Thesis, Aerospace Engineering, Georgia Institute of Technology, Atlanta, Georgia

Fitzpatrick, K.; Feng, Y.; Lind R.; Kurdila, A.J. & Mikolaitis, D.W. (2005). Flow Control in a Driven Cavity Incorporating Excitation Phase Differential. *Journal of Guidance, Control and Dynamics*, Vol. 28, pp.63-70

Gillies, E.A. (1998). Low-Dimensional Control of the Circular Cylinder Wake. *Journal of Fluid Mechanics*, Vol. 371, pp. 157-178

Gracia, M.M. (2010). *Reduced Models to Calculate Stationary Solutions for the Lid-Driven Cavity Problem*. M.Sc. Thesis, Department of Aerospace Science and Technology, Universitat Politècnica de Catalunya, Spain

Gurney, K. (1997). *An Introduction to Neural Networks*. CRC Press

Haykin, S. (1994). *Neural Networks*. Macmillan College Printing Company, New Jersey

Holmes, P.; Lumley, J.L. & Berkooz, G. (1996). *Coherent Structures, Dynamical Systems and Symmetry*. Cambridge University Press, Cambridge, UK

Khataee, A.R.; Zarei, M. & Pourhassan, M. (2010). Bioremediation of Malachite Green from Contaminated Water by Three Microalgae: Neural Network Modeling. *Clean-Soil, Air, Water*, Vol. 38, pp. 96-103

Lieu, T.; Farhat, C. & Lesoinne, M. (2006). Reduced-Order Fluid/Structure Modeling of a Complete Aircraft Configuration. *Computer Methods in Applied Mechanics and Engineering*, Vol. 195, No. 41-43, pp. 5730-5742

Lim, H. & Lee, S. (2002). Flow Control of Circular Cylinders with Longitudinal Grooved Surfaces. *AIAA Journal*, Vol. 40, No. 10, pp. 2027-2036

Lumley, J.L. (1967). The Structure of Inhomogeneous Turbulent Flows. *Atmospheric Turbulence and Radio Propagation*, pp.166-178

Ly, H.V. & Tran, H.T. (2001). Modeling and Control of Physical Processes Using Proper Orthogonal Decomposition. *Mathematical and Computer Modeling*, Vol. 33, pp. 223-236

Mehrotra, K.; Chilukuri, K..M. & Ranka, S. (2000). *Elements of Artificial Neural Networks*. The MIT Press

Newman, A.J. (1996a). *Model Reduction via the Karhunen Loéve Expansion Part 1 An Exposition*. Institute for Systems Research, Technical Report No. 96-32

Newman, A.J. (1996b). *Model Reduction via the Karhunen-Loéve Expansion Part 2: Some Elementary Examples*. Institute for Systems Research, Technical Report No. 96-33

Norberg, C. (1987). Effects of Reynolds Number and a Low-Intensity Free Stream Turbulence on the Flow around a Circular Cylinder. Chalmers University of Technology, ISSN 02809265

Nørgaard, M.; Ravin, O.; Poulsen, N.K. & Hansen, L.K. (2000). *Neural Networks for Modeling and Control of Dynamic Systems*. Springer, London

O'Donnell, B. & Helenbrook, B. (2007). *Proper Orthogonal Decomposition and Incompressible Flow: An Application to Particle Modeling*. Computers and Fluids, Vol. 36, No. 7, pp. 1174-1186

Ong, M.C.; Utnes, T.; Holmedal, L.E.; Myrhaug, D. & Pettersen, B. (2009). Numerical Simulation of Flow around a Smooth Circular Cylinder at Very High Reynolds Numbers. *Journal of Marine Structures*, Vol. 22, No. 2, pp. 142-153

Paksoy, A.; Apacoglu, B. & Aradag, S. (2010). Analysis of Flow over a Circular Cylinder by CFD and Reduced Order Modeling, *Proceedings of ASME 10th Biennial Conference on Engineering Systems Design and Analysis*, Istanbul, Turkey, July 2010

Paksoy, A. & Aradag, S. (2011). Prediction of Lid-Driven Cavity Flow Characteristics Using an Artificial Neural Network Based Methodology Combined With CFD and Proper Orthogonal Decomposition, *Proceedings of the 7th International Conference on Computational Heat and Mass Transfer*, Istanbul, Turkey, July 2011

Samarasinghe S. (2006). *Neural Networks for Applied Sciences and Engineering, From Fundamentals to Complex Pattern Recognition*. Auerbach Publications Taylor and Francis Group

Sanghi, S. & Hasan, N. (2011). Proper Orthogonal Decomposition and Its Applications. *Asia-Pacific Journal of Chemical Engineering*, Vol. 6, pp. 120-128

Siegel, S.; Cohen, K.; Seidel, J.; Aradag, S. & McLaughlin, T. (2008). Low Dimensional Model Development Using Double Proper Orthogonal Decomposition and System Identification, *Proceedings of the 4th Flow Control Conference*, Seattle, Washington, June 2008

Seidel, J.; Cohen, K.; Aradag, S.; Siegel, S. & McLaughlin, T. (2007). Reduced Order Modeling of a Turbulent Three Dimensional Cylinder Wake, *Proceedings of the 37th AIAA Fluid Dynamics Conference and Exhibit*, Miami, Florida, June 2007

Sen., M.; Bhaganagar, K. & Juttijudata, V. (2007). Application of Proper Orthogonal Decomposition (POD) to Investigate a Turbulent Boundary Layer in a Channel with Rough Walls. *Journal of Turbulence*, Vol. 8, No. 41

Sirovich, L. (1987). Turbulence and the Dynamics of Coherent Structures, Part I: Coherent Structures. *Quarterly Applied Mathematics*, Vol. 45, No. 3, pp. 561-571

Smith, T.R.; Moehlis, J. & Holmes P. (2005). Low Dimensional Models for Turbulent Plane Couette Flow in a Minimal Flow Unit. *Journal of Fluid Mechanics*, Vol. 538, pp. 71-110.

Travin, A.; Shur, M.; Strelets, M. & Spalart, P. (1999). Detached-Eddy Simulations Past a Circular Cylinder Flow. *Turbulence and Combustion*, Vol. 63, pp. 293-313

Unal, M. & Rockwell, D. (2002). On Vortex Shedding from a Cylinder, Part 1: The Initial Instability. *Journal of Fluid Mechanics*, Vol. 190, pp. 491-512

Wissink, J.G. & Rodi, W. (2008). Numerical Study of the Near Wake of a Circular Cylinder. *International Journal of Heat and Fluid Flow*, Vol. 29, pp. 1060-1070

Xie G.; Sunden B.; Wang Q. & Tang L. (2009). Performance Predictions of Laminar and Turbulent Heat Transfer and Fluid Flow of Heat Exchangers Having Large TubeRow by Artificial Neural Networks. *International Journal of Heat and Mass Transfer*, Vol. 52, pp. 2484-2497

Zhang, L.; Akiyama, M.; Huang, K.; Sugiyama H. & Ninomiya, N. (1996). Estimation of Flow Patterns by Applying Artificial Neural Networks. *Information Intelligence and Systems*, Vol. 4, pp. 1358-1363

7

Thermal Perturbations in Supersonic Transition

Hong Yan
Northwestern Polytechnical University
P.R.China

1. Introduction

In recent years, there has been considerable interest in the study of bump-based methods to modulate the stability of boundary layers (Breuer & Haritonidis, 1990; Breuer & Landahl, 1990; Fischer & Choudhari, 2004; Gaster et al., 1994; Joslin & Grosch, 1995; Rizzetta & Visbal, 2006; Tumin & Reshotko, 2005; White et al., 2005; Worner et al., 2003). These studies are mostly focused on the incompressible regime and have revealed several interesting aspects of bump modulated flows. Surface roughness can influence the location of laminar-turbulent transition by two potential mechanisms. First, they can convert external large-scale disturbances into small-scale boundary layer perturbations, and become possible sources of receptivity. Second, they may generate new disturbances to stabilize or destabilize the boundary layer. Breuer and Haritonidis (Breuer & Haritonidis, 1990) and Breuer and Landahl (Breuer & Landahl, 1990) performed numerical and experimental simulations to study the transient growth of localized weak and strong disturbances in a laminar boundary layer. They demonstrated that the three-dimensionality in the evolution of localized disturbances may be seen at any stage of the transition process and is not necessarily confined to the nonlinear regime of the flow development. For weak disturbances, the initial evolution of the disturbances resulted in the rapid formation of an inclined shear layer, which was in good agreement with inviscid calculations. For strong disturbances, however, transient growth gives rise to distinct nonlinear effects, and it was found that resulting perturbation depends primarily on the initial distribution of vertical velocity. Gaster *et al.*(Gaster et al., 1994) reported measurements on the velocity field created by a shallow oscillating bump in a boundary layer. They found that the disturbance was entirely confined to the boundary layer, and the spanwise profile of the disturbance field near the bump differed dramatically from that far downstream. Joslin and Grosch (Joslin & Grosch, 1995) performed a Direct Numerical Simulation (DNS) to duplicate the experimental results by Gaster *et al.* (Gaster et al., 1994). Far downstream, the bump generated a pair of counter-rotating streamwise vortices just above the wall and on either side of the bump location, which significantly affected the near-wall flow structures. Worner *et al.* (Worner et al., 2003) numerically studied the effect of a localized hump on Tollmien-Schlichting waves traveling across it in a two-dimensional laminar boundary layer. They observed that the destabilization by a localized hump was much stronger when its height was increased as opposed to its width. Further, a rounded shape was less destabilizing than a rectangular shape.

Researchers have also studied the effect of surface roughness on transient growth. White *et al.*(White et al., 2005) described experiments to explore the receptivity of transient disturbances to surface roughness. The initial disturbances were generated by a spanwise-periodic array of roughness elements. The results indicated that the streamwise flow was decelerated near the protuberances, but that farther downstream the streamwise flow included both accelerated and decelerated regions. Some of the disturbances produced by the spanwise roughness array underwent a period of transient growth. Fischer and Choudhari (Fischer & Choudhari, 2004) presented a numerical study to examine the roughness-induced transient growth in a laminar boundary layer. The results showed that the ratio of roughness size relative to array spacing was a primary control variable in roughness-induced transient growth. Tumin and Reshotko (Tumin & Reshotko, 2005) solved the receptivity of boundary layer flow to a three dimensional hump with the help of an expansion of the linearized solution of the Navier-Stokes equations into the biorthogonal eigenfunction system. They observed that two counter-rotating streamwise vortices behind the hump entrained the high-speed fluid towards the surface boundary layer. Rizzetta and Visbal (Rizzetta & Visbal, 2006) used DNS to study the effect of an array of spanwise periodic cylindrical roughness elements on flow instability. A pair of co-rotating horseshoe vortices was observed, which did not influence the transition process, while the breakdown of an unstable shear layer formed above the element surface played a strong role in the initiation of transition.

Although the effect of physical bumps on flow instabilities has been studied extensively, far fewer studies have explored the impact of thermal bumps. A thermal bump may be particularly effective at supersonic and hypersonic speeds. One approach to introduce the bump is through an electromagnetic discharge in which motion is induced by collisional momentum transfer from charged to neutral particles through the action of a Lorentz force (Adelgren et al., 2005; Enloe et al., 2004; Leonov et al., 2001; Roth et al., 2000; Shang, 2002; Shang et al., 2005). Another approach is through a high-frequency electric discharge (Samimy et al., 2007). Joule heating is a natural outcome of such interactions, and is assumed to be the primary influence of the notional electric discharge plasma employed here to influence flow stability.

For supersonic and hypersonic flows, heat injection for control have considered numerous mechanisms, including DC discharges (Shang et al., 2005), microwave discharges (Leonov et al., 2001) and lasers (Adelgren et al., 2005). Recently however, Samimy *et al.* (Samimy et al., 2007) have employed Localized Arc Filament Plasma Actuators in a fundamentally unsteady manner to influence flow stability. The technique consists of an arc filament initiated between electrodes embedded on the surface to generate rapid (on the time scale of a few microseconds) local heating. Samimy *et al.* (Samimy et al., 2007) employed this method in the control of high speed and high Reynolds number jets. The results showed that forcing the jet with $m = \pm 1$ mode at the preferred column mode frequency provided the maximum mixing enhancement, while significantly reducing the jet potential core length and increasing the jet centerline velocity decay rate beyond the end of the potential core.

Yan *et al.* (Yan et al., 2007; 2008) studied the steady heating effect on a Mach 1.5 laminar boundary layer. Far downstream of the heating, a series of counter-rotating streamwise vortex pairs were observed above the wall on the each side of the heating element. This implies that the main mechanism of the thermal bump displays some degree of similarity to that of the physical bump. This finding motivates the further study on thermal bumps since they have several advantages over physical bumps. These include the ability to switch on and off on-demand, and to pulse at any desired frequency combination. Yan and Gaitonde (Yan & Gaitonde, 2010) studied both the steady and pulsed thermal rectangular bumps in supersonic boundary layer. For the steady bump, the velocity fluctuation profile across the span bore some similarity to the physical bump in an overall sense. The disturbance decayed downstream, suggesting that the linear stability theory applies. For pulsed heating, non-linear dynamic vortex interactions caused disturbances to grow dramatically downstream. Yan and Gaitonde (Yan & Gaitonde, 2011) assessed the effect of the geometry of the thermal bump and the pulsing properties. It was shown that the rectangular element was more effective than the circular counterpart. The smaller width of the rectangular element produced higher disturbance energy, while the full-span heating indicated delayed growth of the disturbances. The disturbance energy increased with the initial temperature variation, and the lower frequency produced lesser disturbance energy.

This chapter summarizes some of the key findings in thermal perturbation induced supersonic flow transition in our research group. The chapter is organized as follows. The flow configuration, setup and numerical components are described first. The effect of the pulsed heating is then explored in the context of disturbance energy growth, and correlated with linear stability analysis. Subsequently, the phenomenology of the non-linear dynamics between the vortices produced by the pulsed bump and the compressible boundary layer is examined with emphasis on later stages of the boundary layer transition.

2. Flow configuration

A Mach 1.5 flat plate flow is considered with the total temperature and pressure of 325 K and 3.7×10^5 Pa, respectively. The thermal bump is imposed as a surface heating element and centered in the spanwise direction as shown schematically in Fig. 1. For some simulations, the

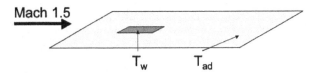

Fig. 1. Flat plate with thermal bump

nominally two-dimensional case is considered where the bump extends cross the entire span of the plate. Even for this case, the three-dimensional domain is discretized since the primary disturbance growth is three-dimensional. The heating effect is modeled as a time-dependent step surface temperature rise ΔT_w with a monochromatic pulsing frequency (f) and duty cycle as shown in Fig. 2, where the pulsing time period $T_t = 1/f$. The subscript w denotes the value at the wall. For simplicity, it is assumed that $\Delta T_w = T_w - T_{w0}$, where T_w and T_{w0} are wall temperature inside and outside of the heating region, respectively, and T_{w0} is fixed at the adiabatic temperature (T_{ad}) as shown in Fig. 1.

In all perturbed simulations, the heating element is placed immediately upstream of the first neutral point in the stability neutral curve for an adiabatic flat plate boundary layer with the freestream Mach number (M_∞) of 1.5. The stability diagram, shown in Fig. 3, is obtained from the Langley Stability and Transition Analysis Codes (LASTRAC) (Chang, 2004). LASTRAC performs linear calculations and transition correlation by using the N-factor method based on linear stability theory, where the N factor is defined by $N = \int_{s_0}^{s_1} \gamma ds$, and s_0 is the point at which the disturbance first begins to grow, s_1 is the end point of the integration, which may be at upstream or downstream of where transition is correlated and γ is the characteristic growth rate of the disturbance. For disturbances at $f = 100$ kHz, the first neutral point is located at the Reynolds number of $Re = 610$ based on the similarity boundary-layer length scale (η) defined as $= \sqrt{\nu_\infty x / u_\infty}$, where ν_∞ and u_∞ are the freestream kinematic viscosity

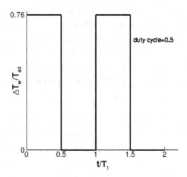

Fig. 2. Two time periods of surface temperature rise, T_t

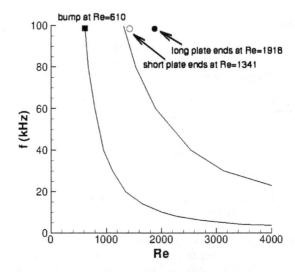

Fig. 3. Neutral curve for Mach 1.5 adiabatic flat plate boundary layer

and streamwise velocity, respectively and is shown as the solid rectangle in Fig. 3. The local Reynolds number based on the running distance from the leading edge of the plate[1] is defined by $Re_x = Re^2$. Thus, the distance from the leading edge of the plate to the leading edge of the heating element is 7.65 mm (i.e. corresponding to $Re = 610$). The nominal spanwise distance between bumps is determined from the most unstable mode, which for the present Mach number is oblique. The N factor profile, shown in Fig. 4 for different spanwise wave lengths (λ) at $M_\infty = 1.5$ and $f = 100$ kHz, indicates that $\lambda = 2$ mm is the most unstable mode. Thus, the nominal distance between two adjacent heating elements is set to 2 mm to excite the most unstable wave. This is accomplished by choosing a spanwise periodic condition on a domain of 2 mm, at the center of which a bump is enforced.

---

[1] Re_x grows linearly with x and is adopted in all the figures and tables except in the neutral stability curve figure, where Re is used instead.

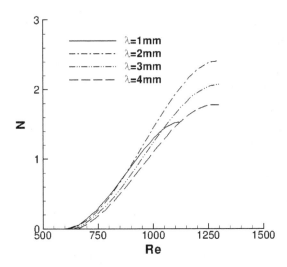

Fig. 4. N factor for different spanwise wavelengths at Mach=1.5 and $f = 100$ kHz

3. Numerical model

The governing equations are the full compressible 3-D Navier-Stokes equations. The Roe scheme (Roe, 1981) is employed together with the Monotone Upstream-centered Schemes for Conservation Laws (MUSCL)(Van Leer, 1979) to obtain up to nominal third order accuracy in space. Solution monotonicity is imposed with the harmonic limiter described by Van Leer (Van Leer, 1979). Given the stringent time-step-size limitation of explicit schemes, an implicit approximately factored second-order time integration method with a sub-iteration strategy is implemented to reduce computing cost. The time step is fixed at 4.2×10^{-8} s for all the cases.

The Cartesian coordinate system is adopted with x, y and z in the streamwise, wall-normal and spanwise direction, respectively. The x axis is placed through the center of the plate with the origin placed at the leading edge of the plate. The computational domain is L_x =38 mm long, L_y =20 mm high and L_z =2 mm wide for case 1, and L_x =76 mm long, L_y =51 mm high and Lz =2 mm wide for cases 2 and 3. This is determined by taking two factors into consideration. In the streamwise direction, the domain is long enough to capture three-dimensional effects induced by heating and to eliminate the non-physical effects at the outflow boundary. Based on this constraint, the Reynolds number at the trailing edge of the plate is $Re_L = 1.80 \times 10^6$ for case 1, and 3.68×10^6 for cases 2 and 3. In the wall-normal direction, the domain is high enough to avoid the reflection of leading edge shock onto the surface. The upper boundary is positioned at $86\delta_L$ above the wall for case 1, and $220\delta_L$ for cases 2 and 3, where δ_L is the boundary layer thickness at the trailing edge of the plate. The velocity, pressure and density in the figures shown in Section *Results and analyses* are normalized by u_∞, p_∞ and ρ_∞, respectively. The vorticity is normalized by u_∞/L_x, where $u_\infty = 450$ m/s and L_x =0.038 m.

The grid is refined near the leading edge of the flat plate and near the heating element. Approximately 150 grid points are employed inside the boundary layer at the leading edge of the heating element to resolve the viscous layer and capture the heat release process. Previous results (Yan & Gaitonde, 2008) indicated that this is fine enough to capture the near-field effect of the thermal perturbation. The grid sizes are $477 \times 277 \times 81$ in the x, y and z direction, respectively for case 1, and $854 \times 297 \times 81$ for cases 2 and 3.

For boundary conditions, the stagnation temperature and pressure and Mach number are fixed at the inflow. The no-slip condition with a fixed wall temperature is used on the wall. The pulse is imposed as a sudden jump as shown in Fig. 2. The symmetry condition is enforced at the spanwise boundary to simulate spanwise periodic series of heating elements spaced L_z apart in the finite-span bump cases. This boundary condition is also suitable to mimic two-dimensional perturbation in the full-span bump case. First-order extrapolation is applied at the outflow and upper boundaries.

4. Results and analyses

The study is comprised of two parts. The first part studies the effects of the pulsed bump whose properties are listed in case 1 in Table 1. The pulsed bump introduces the disturbance at $f = 100$ kHz, and is located at $Re_0 = 610^2 = 0.3721 \times 10^6$, immediately upstream of the first neutral point (Re=610) for this particular frequency, where the subscript 0 denotes the streamwise location of the thermal bump. The Reynolds number at the trailing edge of the plate is $Re_L = 1341^2 = 1.80 \times 10^6$, which corresponds to the location immediately downstream of the second neutral point (Re =1300) as shown in Fig. 3. The rectangular bump is considered with spanwise width (w) of 1 mm and its streamwise length (l) is arbitrarily set to 0.2 mm.

Case	w/l (mm)	f (kHz)	$Re_L \times 10^{-6}$
1	1/0.2	100	1.80
2	1/0.2	100	3.68
3	2/0.2 (full span)	100	3.68

Table 1. Classification of cases simulated, $\Delta T_w = 0.76 T_{ad}$, $Re_0 = 0.3721 \times 10^6$

The second part examines the breakdown process at later stages of flow evolvement. To this end, the plate is extended far downstream of the second neutral point to $Re = 1918$ ($Re_L = 3.68 \times 10^6$) as shown in Fig. 3. Both 3D and 2D thermal bumps are considered. The cases are denoted as cases 2 and 3 in Table 1.

4.1 Effect of pulsed bump

4.1.1 Unperturbed flow (basic state)

The basic or unperturbed state is a Mach 1.5 adiabatic flat plate boundary layer with Reynolds number at the trailing edge of the plate of $Re_L = 1.80 \times 10^6$. Figs. 5 and 6 show the streamwise and wall-normal velocity profiles along the y direction at $Re_x = 1.4 \times 10^6$ at the spanwise center and side of the plate. The y coordinate is normalized with the local theoretical boundary layer thickness (δ). Both boundary layer thickness and velocity profiles are predicted correctly compared to the compressible boundary layer theory. In particular, the wall-normal velocity, which is of much smaller order $v \sim u_\infty / \sqrt{Re_x}$, is captured correctly as well. The fact that the profiles on the center and side of the plate collapse demonstrates flow two-dimensionality as expected.

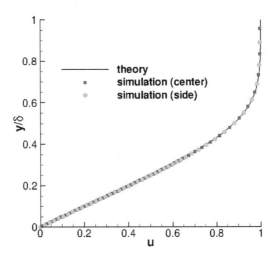

Fig. 5. Streamwise velocity in the y direction at $Re_x = 1.4 \times 10^6$ (basic state for case 1)

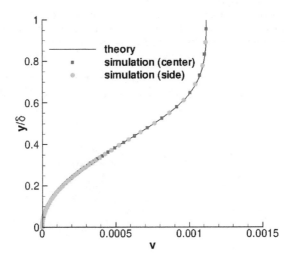

Fig. 6. Vertical velocity in the y direction at $Re_x = 1.4 \times 10^6$ (basic state for case 1)

4.1.2 Perturbed flow by pulsed bump

A pulsed thermal bump at a frequency of 100 kHz is turned on to introduce the disturbance after the basic state is obtained. Recall that the bump is placed immediately upstream of the first neutral point (where $Re = 610$) for disturbances at frequency of 100 kHz.

For all the pulsed heating cases, the solution is marched until a statistically stationary state is obtained. This determination is made by monitoring all primitive variables at several points in the domain. Mean statistics are then gathered over numerous cycles until time invariant values are obtained. The instantaneous results presented are those obtained after the time-mean quantities reach invariant values.

Fig. 7 shows the instantaneous streamwise vorticity contours on the wall. Since these values

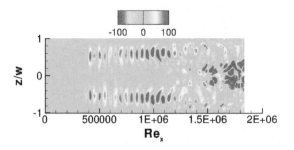

Fig. 7. Instantaneous ω_x contours on the wall (case 1)

are plotted after the flow reaches an asymptotic state, the vortex pattern is formed by the dynamic vortex interaction from numerous heating pulses. When the bump is pulsed, a complex vortex shedding and dynamic interaction process results in a vortical pattern with the alternating sign in the streamwise direction. These structures are constrained in the spanwise direction, but move away from the surface, which will be shown in the time-mean values. Smaller eddies are observed at about $Re_x = 1.25 \times 10^6$ near the central region and intensified downstream of $Re_x = 1.5 \times 10^6$.

The effect of pulsing on the time-mean streamwise vorticity is shown in Fig. 8. The spanwise-varying streaky structures are formed downstream with concentration in the central

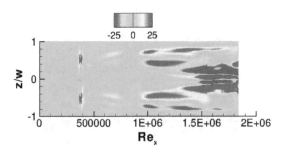

Fig. 8. Time-mean ω_x contours on the wall (case 1)

region and intensified after $Re_x = 1.5 \times 10^6$. These results bear some similarity to free shear flow control with tabs. For example, Zaman et al. (Zaman et al., 1994) demonstrated with a comprehensive experimental study that the pressure variation induced by the tabs installed on the nozzle wall generated streamwise vorticity, which significantly enhanced the mixing downstream of the nozzle exit.

The vortex interaction and penetration can be seen on the cross sections in Fig. 9. The first cross section (Fig. 9(a)) is cut immediately downstream of the bump, therefore the top pair

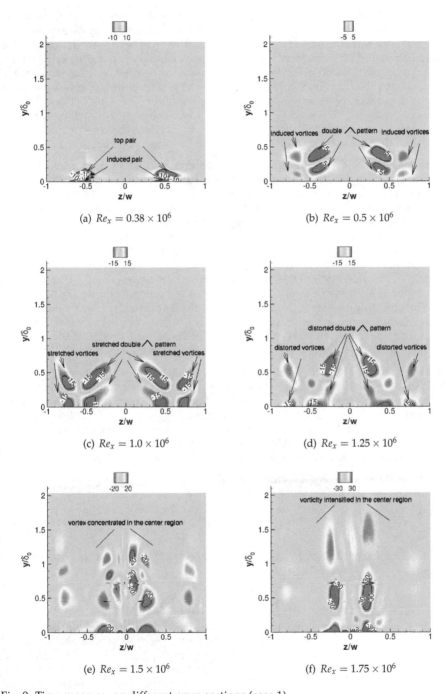

Fig. 9. Time-mean ω_x on different cross sections (case 1)

of vortices above the wall possesses the same sign as that at the leading edge of the bump shown in Fig. 8 (positive at $z = 0.5w$ and negative at $z = -0.5w$). As they move downstream, they are lifted away from the wall and induce additional vortices near the wall to satisfy the noslip condition as well on the sides where periodic conditions apply. The original pairs form a double $\wedge$ pattern as indicated in Fig. 9(b). As they move downstream, vortices are stretched and intensified as shown in Fig. 9(c). At $Re_x = 1.25 \times 10^6$ (Fig. 9(d)), the vortex pattern is distorted and the vortices break into smaller eddies. This leads to the complex vortex dynamic interaction downstream at $Re_x = 1.5 \times 10^6$ (Fig. 9(e)), which completely distorts the double $\wedge$ pattern and results in a vortex trace that appears to move towards the center region. At the last station (Fig. 9(f)), the vorticity is intensified around the center region.

The accumulated effect of the streamwise vorticity distorts the basic state in nonlinear fashion. Fig. 10 shows the streamwise perturbation velocity contours on the downstream cross sections. The quantity plotted is $\bar{u} - u_b$, where $\bar{u}$ is the time-mean value of the pulsed case. Please note change in contour levels between Figs. 9 and 10. Immediately downstream of the bump (Fig. 10(a)), a velocity excess region is formed above the wall due to flow expansion. Proceeding downstream, a velocity deficit is generated downstream of the center of the heating element, while an excess is observed on both sides of the bump (Fig. 10(b)). This behavior is similar to the observation in the flow over a shallow bump studied by Joslin and Grosch (Joslin & Grosch, 1995) and the steady heating case discussed earlier. The intensity of the excess region is at the same level as that in the steady heating (compare Fig. 10(a) with Fig. 9(a)). Proceeding downstream, the pulsed bump behaves differently from the steady one. The velocity distortion is amplified as seen in Figs. 10(c) and (d). The velocity excess regions grow in the region near the wall across the entire span of the domain (Figs. 10(e) and (f)) as the streamwise vortices bring the high-momentum fluid from the freestream towards the wall.

The above observations are further explored in Fig. 11, which plots $\bar{u}$ and $u' = \bar{u} - u_b$ along the y direction at $z=0$ and $z=-0.5w$ (i.e., at the spanwise edge of the bump). The intensity of the velocity excess in the near-wall region increases along the downstream and reaches about 20% of u_∞ at $Re_x = 1.75 \times 10^6$, while in the outer region, a velocity deficit is observed. This results in an inflection point in the mean flow near the centerline (Fig. 11(a)), giving rise to the rapid breakdown observed in Fig. 9. On the edges of the bump, the flow is accelerated cross the entire boundary layer and no inflection points are observed (Fig. 11(b)).

The strength of disturbance energy growth for the compressible flow is measured by the energy norm proposed by Tumin and Reshotko (Tumin & Reshotko, 2001) as

$$E = \int_0^\infty \vec{q}^T A \vec{q} \, dy \tag{1}$$

where $\vec{q}$ and A are the perturbation amplitude vector and diagonal matrix, respectively, and are expressed as

$$\vec{q} = (u', v', w', \rho', T')^T \tag{2}$$

$$A = diag[\rho, \rho, \rho, T/(\gamma \rho M_\infty^2), \rho/(\gamma(\gamma - 1) T M_\infty^2)] \tag{3}$$

The first three terms represent the components of the kinetic disturbance energy denoted as E_u, E_v and E_w, respectively and the last two represent the thermodynamic disturbance energy as E_ρ and E_T. The spanwise-averaged disturbance energy is plotted in Fig. 12. The initial growth rate of the total disturbance energy is small and becomes larger as the disturbances are amplified in the region of $0.9 \times 10^6 < Re_x < 1.4 \times 10^6$. The disturbances then saturate and reach finite amplitude shown as a plateau in Fig. 12(a). At this stage, the flow reaches

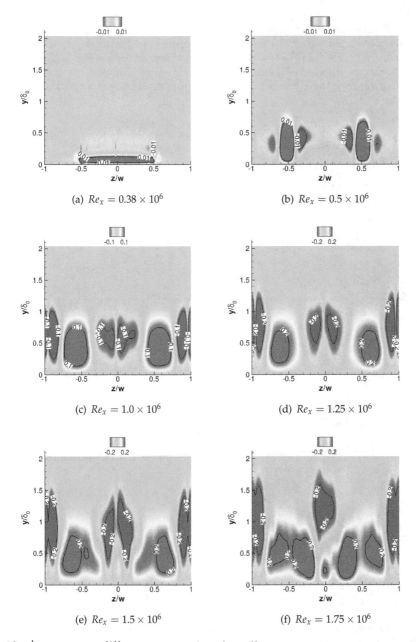

(a) $Re_x = 0.38 \times 10^6$

(b) $Re_x = 0.5 \times 10^6$

(c) $Re_x = 1.0 \times 10^6$

(d) $Re_x = 1.25 \times 10^6$

(e) $Re_x = 1.5 \times 10^6$

(f) $Re_x = 1.75 \times 10^6$

Fig. 10. $u' = \bar{u} - u_b$ on different cross sections (case 1)

a new state which becomes a basic state on which secondary instabilities can grow (Schmid & Henningson, 2001). The new basic state is represented by the appearance of the inflection point at $Re_x = 1.5 \times 10^6$ in the left plot of Fig. 11(a). The disturbances grow more rapidly after $Re_x = 1.5 \times 10^6$ because the secondary instability susceptible to high frequency disturbances usually grows more rapidly than the primary instabilities. The thermodynamic disturbance energy (E_ρ and E_T) in Fig. 12 (b) shows a similar trend except for a spike in the vicinity of the thermal bump as expected. However the thermodynamic components are four orders of magnitude lower than the E_u, indicating that the primary disturbance quickly develops a vortical nature.

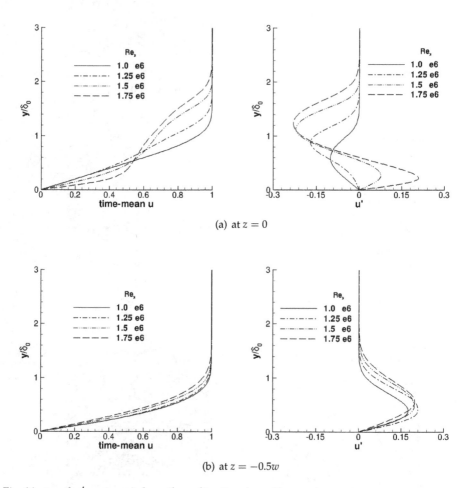

(a) at $z = 0$

(b) at $z = -0.5w$

Fig. 11. $\bar{u}$ and $u' = \bar{u} - u_b$ along the y direction (case 1)

With pulsed heating, the disturbances grow significantly downstream and the flow becomes highly inflectional. The observation is consistent with the linear stability theory. However the velocity fluctuation reaches 20% of u_∞ at the downstream, indicating that non-linear growth comes into the play and the assumption that disturbances are infinitesimal is not valid any

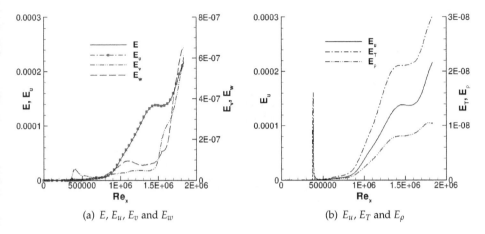

(a) E, E_u, E_v and E_w (b) E_u, E_T and E_ρ

Fig. 12. Spanwise-averaged disturbance energy along the x direction (case 1)

more. Thus, the dynamic vortex non-linear interaction plays an important role in the later development of the sustained disturbance growth, and will be discussed in the following section.

4.2 Analyses of breakdown process

This section explores the phenomenology of the non-linear dynamics between the vortices produced by the bump and the compressible boundary layer. To this end, the domain size is extended in both streamwise and normal directions relative to the case 1, but the spanwise width remains unchanged. The Reynolds number at the end of the plate is $Re_L = 3.68 \times 10^6$. Two cases (cases 2 and 3 in Table 1) are examined; the first considers a three-dimensional perturbation associated with a finite-span thermal bump, and the second is comprised of full-span disturbances. In both cases, the bump is positioned at the same streamwise location as in the case 1 with the same pulsing frequency and magnitude shown in Table 1.

A new basic state (no perturbation) is obtained for the cases with the extended domain. In the absence of imposed perturbations, no tendency is observed towards transition even at the higher Reynolds number. Figs. 13 and 14 show the streamwise and wall-normal velocity profiles along the y direction at $Re_x = 3.5 \times 10^6$. The comparisons with the compressible theoretical profiles are good and the fact that the profiles on the center and side of the plate collapse demonstrates flow two-dimensionality as expected.

The heating element is turned on after the basic state is obtained. For the finite-span case, a series of counter-rotating streamwise vortices are generated at the edges of the thermal bump by heating induced surface pressure variation as discussed earlier. These vortices shed from their origins when the element is switched off, forming a traveling vortical pattern with an alternating sign in the streamwise direction up to $Re_x = 1.25 \times 10^6$ as shown in Fig. 15(a), where the instantaneous streamwise vorticity contours are plotted on the wall. Further downstream, small organized alternating structures appear near the center region at $Re_x = 1.5 \times 10^6$. Up to this point, the perturbed flow structures are similar to case 1 as expected. Subsequently, the vortices are intensified at about $Re_x = 2.0 \times 10^6$ due to vortex stretching

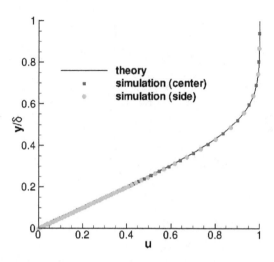

Fig. 13. Streamwise velocity in the y direction at $Re_x = 3.5 \times 10^6$ (basic state for cases 2 and 3)

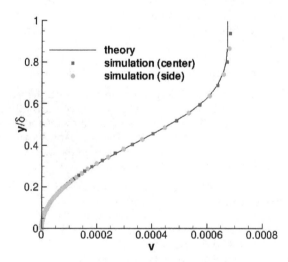

Fig. 14. Vertical velocity in the y direction at $Re_x = 3.5 \times 10^6$ (basic state for cases 2 and 3)

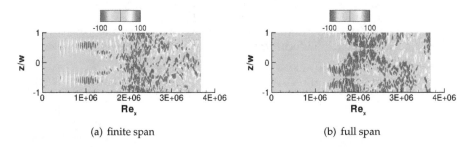

(a) finite span   (b) full span

Fig. 15. Instantaneous ω_x contours on the wall (cases 2 and 3)

and interaction, which is described in more detail later. After $Re_x = 2.5 \times 10^6$, the flow tends to relax to a relatively universal stage. The full-span case shows a different development as shown in Fig. 15(b). The counter-rotating vortices are not observed immediately downstream of the bump. Rather, an asymmetric instantaneous vortical pattern is initiated with small successive structures starting at about $Re_x = 1.25 \times 10^6$, which are concentrated on the lower half of the domain. The fact that these small structures occur at the same location for both cases suggests that they are unlikely to be related to the original counter-rotating vortices, and an inherent stability mechanism that stimulates their appearance.

The vortex development is examined in a three-dimensional fashion in Fig. 16 which shows the iso-surface of the non-dimensionalized vorticity magnitude at $|\omega| = 100$ colored with

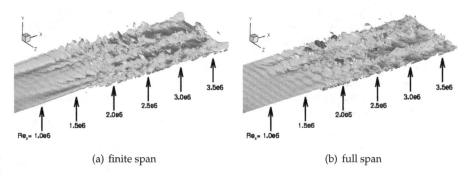

(a) finite span   (b) full span

Fig. 16. Iso-surface of instantaneous vorticity magnitude at $|\omega| = 100$, $x : y = 1 : 10$, $x : z = 1 : 10$ (cases 2 and 3)

the distance from the wall. This iso-level is chosen to reveal the near-wall structures. For visualization purpose, the y and z axes are equally stretched with a ratio to the x axis of 10. The same length unit is used for these three axes and Reynolds numbers are only marked for discussion purpose. Thus the structures are closer to the wall and more elongated in the streamwise direction than they appear in the plot. In the finite-span case shown in Fig. 16(a), a sheet of vorticity is generated by the wall shear and rolls up into three rows of hairpin-like vortices across the span at about $Re_x = 1.5 \times 10^6$. The vortices are then slightly lifted away from the wall at about $Re_x = 2.0 \times 10^6$. Correspondingly, the vortices are stretched in the streamwise direction in the wall region, resulting in the intensification of streamwise vorticity as previously shown in Fig. 15(a). The vortices become weaker as the flow relaxes further downstream. As shown in Fig. 16(b), hairpin-like structures are also observed for the full-span case, but they develop in an asymmetric fashion. Similar to the full-span case, vortex stretching in the lifting process induces strong streamwise vorticity in the wall region.

The hairpin structures are better displayed by lowering the iso-levels to $|\omega| = 25$ as shown in Fig. 17. The hairpin-like vortices are initiated across the span at about $Re_x = 2.0 \times 10^6$.

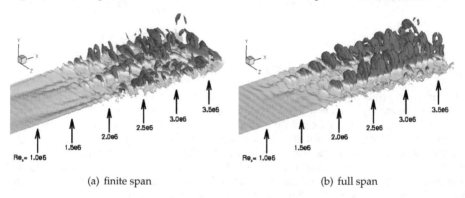

(a) finite span   (b) full span

Fig. 17. Iso-surface of instantaneous vorticity magnitude at $|\omega| = 25$, $x : y = 1 : 10$, $x : z = 1 : 10$ (cases 2 and 3)

The legs of the hairpin constitute a pair of counter-rotating vortices oriented in the streamwise direction in the wall region. They are mainly comprised of ω_x and can be difficult to discern in the total vorticity iso-surface plot since the spanwise vorticity ω_z is dominant in the boundary layer. On the other hand, the heads, mainly comprised of ω_z, can be easily identified in the total vorticity variable because they penetrate into the boundary layer about 2.24δ and 2.35δ at $Re_x = 2.5 \times 10^6$ and 3.5×10^5, respectively, where δ is the local unperturbed laminar boundary layer thickness. It is also observed that for the full-span case, the hairpin vortices are tilted higher in the boundary layer than in the finite-span case. The fact that hairpin vortices appear in the full-span case confirms that the initial counter-rotating streamwise vortices are not a necessity in generating the hairpin vortices.

The vorticity concentration can be viewed through vorticity deviation from the basic state as shown in Fig. 18. Looking downstream, close examination reveals that the right leg rotates with positive ω_x and the head with negative ω_z. Three hairpin vortices are annotated on the plot. The legs can be more clearly seen in the iso-surface of ω_x difference in Fig. 19 and the head in the iso-surface of ω_z difference in Fig. 20. Since the value of ω changes, the structures appear to be broken, but other values confirm the coherence of the structures. The hairpin vortices are aligned in the streamwise direction, forming a pattern similar to K-type breakdown, which results from fundamental modes (Klebanoff et al., 1962). In addition, they appear to be highly asymmetric for both cases. Robinson (Robinson, 1991) pointed out that in a turbulent boundary layer, the symmetry of vortex was predominantly distorted, yielding structures designated "one-legged hairpins". Fig. 21 shows a hairpin vortex schematically. Low-momentum fluid is lifted away from the wall between the legs while high-momentum fluid from the freestream is brought down to the wall outside the legs.

The above described motion of the hairpin vortices alters the velocity distribution in the wall region. In the finite-span case, the passage of the counter-rotating vortices generates

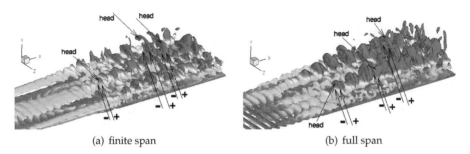

(a) finite span   (b) full span

Fig. 18. Iso-surface of instantaneous vorticity magnitude difference at $|\omega| - (|\omega|)_b = 25$, $x : y = 1 : 10$, $x : z = 1 : 10$ (cases 2 and 3)

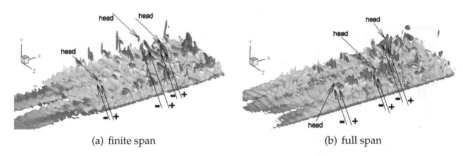

(a) finite span   (b) full span

Fig. 19. Iso-surface of instantaneous streamwise vorticity difference at $\omega_x - (\omega_x)_b = \pm 15$, $x : y = 1 : 10$, $x : z = 1 : 10$ (cases 2 and 3)

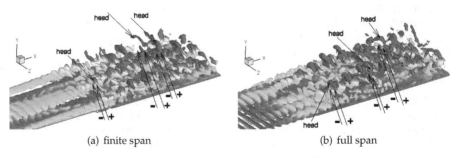

(a) finite span   (b) full span

Fig. 20. Iso-surface of instantaneous spanwise vorticity difference $\omega_z - (\omega_z)_b = 25$, $x : y = 1 : 10$, $x : z = 1 : 10$ (cases 2 and 3)

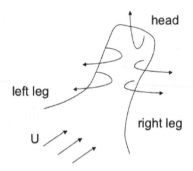

Fig. 21. Schematic of hairpin vortex in pulsed heating

several streamwise streaks in the center region and a low-speed streak is flanked alternately by high and low-speed streaks as shown in Fig. 22(a), which plots the instantaneous streamwise

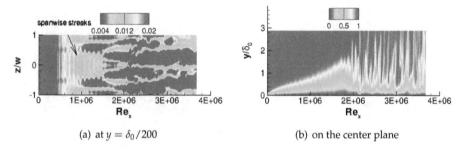

(a) at $y = \delta_0/200$ (b) on the center plane

Fig. 22. Instantaneous u contours for the finite-span case (case 2)

velocity (u) contours at the first grid point above the wall (*i.e.* $y = \delta_0/200$). The central low-speed region is intensified and concentrated towards the center between $1.5 \times 10^6 < Re_x < 2.0 \times 10^6$, resulting in a strong growth of the boundary layer as shown in Fig. 22(b), which depicts the velocity contours on the symmetry plane passing through the center of the domain. At $Re_x = 2.0 \times 10^6$ which is just downstream of the second neutral point, the flow pattern changes dramatically. The low-speed streaks weaken in the wall region and the near-wall low-momentum region becomes thinner as the hairpin vortices pump the low-momentum fluid away from the wall. Strong three-dimensional fluctuations are observed in the upper portion of the boundary layer where the hairpin vortices interact with the high-momentum fluid, leading to the boundary layer distortion.

In the full-span case, the initial spanwise structures are almost two-dimensional in nature. Subsequently, the low-speed streaks are formed at about the location where the hairpin vortices start to appear as shown in Fig. 23(a). This indicates that the low-speed streaks are the footprints of the hairpin vortices. The boundary layer growth is not as strong as that in the finite-span case between $1.5 \times 10^6 < Re_x < 2.0 \times 10^6$ (compare Figs. 23(b) and 22(b)). However, downstream of $Re_x = 2.0 \times 10^6$, strong three-dimensional fluctuations are observed, similar to the finite-span case. It suggests that the non-linear disturbance growth becomes dominant and the initial disturbance form becomes less important. This

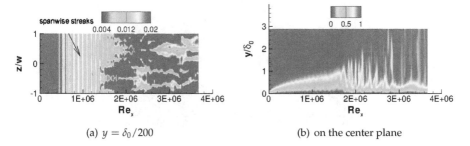

(a) $y = \delta_0/200$ (b) on the center plane

Fig. 23. Instantaneous u contours for the full-span case (case 3)

is confirmed in the disturbance energy growth in Fig. 24, which plots the spanwise-averaged

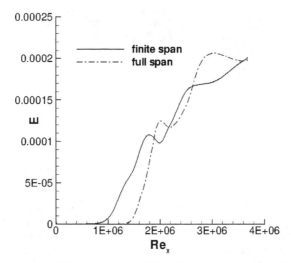

Fig. 24. Spanwise-averaged time-mean total disturbance energy along the x direction (cases 2 and 3)

time-mean total disturbance energy for both finite- and full-span cases. The energy growth in the 2-D perturbations is much weaker than that in the 3-D ones near the bump. However, as the non-linear stability mechanism becomes dominant after about $Re_x = 2.0 \times 10^6$, the disturbance energy growth in both cases becomes comparable.

The accumulated effect of high-frequency pulsing is now described by the time-mean quantities. Only the finite-span results are shown unless otherwise specified. The time-mean pressure ($\bar{p}$) contours are shown on the center plane in Fig. 25. A series of expansion waves is formed at about $Re_x = 2.0 \times 10^6$ and propagates outside the boundary layer. This is caused by the strong boundary layer distortion as shown in the time-mean streamwse velocity contours on the center plane in Fig. 26. The momentum thickness at $Re_x = 2.0 \times 10^6$ is increased by a factor of 1.7 compared to that in laminar flow, indicating that the boundary layer is highly energized downstream of $Re_x = 2.0 \times 10^6$ and shows signs of transition to turbulence. The

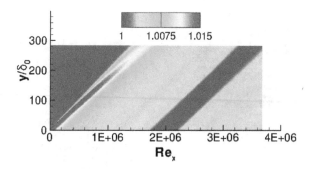

Fig. 25. $\bar{p}$ contours on the center plane for the finite-span case (case 2)

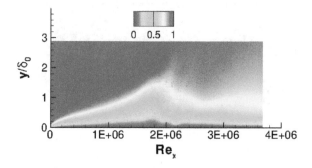

Fig. 26. $\bar{u}$ contours on the center plane for the finite-span case (case 2)

expansion waves in the downstream location are also partially observed in case 1, in which the outlet boundary is set at $Re_L = 1.80 \times 10^6$.

The boundary layer distortion can be assessed by the variation of shape factor H obtained from the mean velocity profile as shown in Fig. 27. The shape factor for the basic state, shown for comparison, reaches an asymptotic value of 2.6 as the flow becomes fully-developed laminar (Fig. 27(a)). With heating, the mean flow is strongly distorted, causing the shape factor to oscillate taking values of 2.85 and 1.35 between $Re_x = 1.5 \times 10^6$ and 2.0×10^6, respectively as shown in Fig. 27(a). A lower shape factor indicates a fuller velocity profile. After $Re_x = 2.0 \times 10^6$ the shape factor decreases rapidly, indicating an increase of the flow momentum in the boundary layer, and starts to level off around $Re_x = 3.0 \times 10^6$. Strong spanwise non-uniformity is observed at $Re_x = 1.5 \times 10^6$ and 2.0×10^6 as shown in Fig. 27(b), while in later stages, only mild distortion is observed and the shape factor reduces to around 1.5, which is close to the turbulent value.

Features of the turbulence statistics are examined through the transformed velocity and Reynolds stresses. Fig. 28 shows the transformed velocity profiles at different downstream locations along the center line ($z=0$) and the side line of the bump ($z=-0.5w$). In the viscous sublayer of a compressible turbulent boundary layer where $y^+ < 5$, the turbulent stresses are negligible compared to viscous stress and the velocity near the wall grows linearly with the distance from the wall as $u^+ = y^+$, where u^+ is defined as u_{vd}/u_τ, and y^+ as $y u_\tau / v_w$. The friction velocity u_τ is defined as $\sqrt{\tau_w/\rho_w}$, where τ_w is wall stress. The detailed formulation of the transformed velocity u_{vd} may be found in Smits and Dussauge (Smits & Dussauge, 2006).

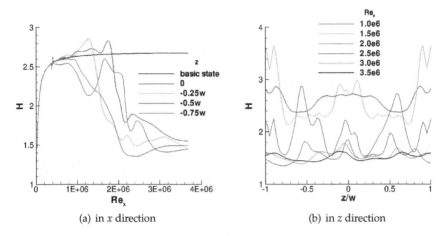

(a) in x direction   (b) in z direction

Fig. 27. Shape factor for the finite-span case (case 2)

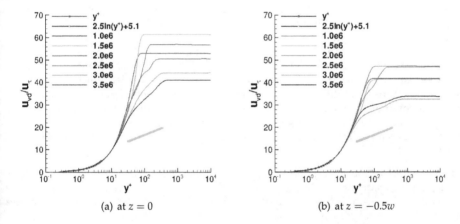

(a) at $z = 0$ (b) at $z = -0.5w$

Fig. 28. Transformed velocity for the finite-span case at different Reynolds number (case 2)

Good agreement is found with the theory at different Reynolds numbers at both center and side locations. The turbulent stresses become large between $y^+ > 30$ and $y/\delta \ll 1$ where the log law holds with $u^+ = \frac{1}{\kappa}\ln(y^+) + C$ with $\kappa = 0.4$ and $C = 5.1$ (Smits & Dussauge, 2006). It is shown in Fig. 28 that the logarithmic region gradually forms with increasing Reynolds number and the velocity slope approaches the log law. However, a large discrepancy remains between the velocity profile at the end of the plate ($Re_x = 3.5 \times 10^6$) and the log law, indicating that the perturbed flow has not reached fully-developed turbulence.

The Reynolds stress profiles are shown to further examine the evolution of the flow. Fig. 29 shows the streamwise Reynolds stress ($\overline{\rho u'u'}$) and Reynolds shear stress ($\overline{\rho u'v'}$) normalized by the local wall stress (τ_w) at $Re_x = 3.5 \times 10^6$. Note that the local boundary layer thickness δ varies across the span. Experimental and numerical results by other researchers (Johnson

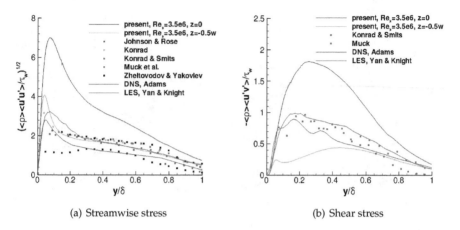

(a) Streamwise stress   (b) Shear stress

Fig. 29. Reynolds stress for the finite-span case (case 2)

& Rose, 1975; Konrad, 1993; Konrad & Smits, 1998; Muck et al., 1984; Yan et al., 2002; Zheltovodov et al., 1990) are plotted for comparison. The predicted streamwise Reynolds stress presents a similar trend to the experiments and other numerical data. It reaches the peak at about $y = 0.05\delta$–0.1δ and decays rapidly between $0.1\delta < y < 0.3\delta$. A large spanwise variation of the peak value is observed with the value at $z = 0$ being 1.8 times that at $z = -0.5w$. The same observation holds for the Reynolds shear stress as shown in Fig. 29(b), which is a main source of turbulence production in the wall-bounded flows. The largely scattered data implies that the flow is still in transitional stage, where the strong non-linear disturbances continue to extract energy from the mean flow to maintain their mobility before the energy redistribution equilibrates and the flow exhibits some features of fully-developed turbulence. This is also consistent with that the mean velocity profile being located above the log law in Fig. 28.

Overall, the effect of the disturbance introduced by thermal bumps is observed to follow classical stability theory in the linear growth region. For the parameters considered, the gross features of transitional flow appear near the second neutral point. These features consist of hairpin vortex structures which are non-staggered and resemble K-type transition. Comparison of 3-D (finite span) with 2-D (full span) perturbations effects indicate that although the near field consequences of the bump are profoundly different, the development further downstream is relatively similar, suggesting a common non-linear mechanism associated with the interaction of the disturbance with the boundary layer vorticity.

5. Concluding remarks

This chapter explores the stability mechanism of a thermally perturbed Mach 1.5 flat plate boundary layer. With pulsed heating at frequency of 100 kHz immediately upstream of the first neutral point, non-linear dynamic vortex interactions cause disturbances to grow dramatically downstream and the maximum velocity fluctuation reaches about 20% of u_∞. The inflectional velocity profile makes the flow highly susceptible to the secondary instabilities.

The dynamic vortex interaction at later stages of the boundary layer development is studied by extending the flat plate further downstream. Hairpin structures, considered as one kind of the basic structures in turbulence, are observed and serve to increase the momentum in the wall region. The fact that the hairpin vortices are observed in the full-span case suggests that the initial counter-rotating vortices generated by the finite-span bump might not be directly associated with the formation of hairpin structures. The boundary layer is observed to grow noticeably downstream relative to the unperturbed case. The Reynolds stresses and shape factor profiles suggest that the boundary layer is approaching turbulence, but remains transitional at the end of the computational domain. These results suggest that pulsed heating can be used as an effective mechanism to modulate the supersonic laminar-turbulence transition. One effective way to generate pulsed heating is through plasma actuator where Joule heating and electrode heating are effectively assumed as surface heating.

6. References

Adelgren, R., Yan, H., Elliott, G., Knight, D., Beutner, T. & Zheltovodov, A. (2005). Control of edney iv interaction by pulsed laser energy deposition, *AIAA J.* 43(2): 256–269.

Breuer, K. S. & Haritonidis, J. H. (1990). The evolution of a localized disturbance in a laminar boundary layer. part i. weak disturbances, *J. Fluid Mech.* 220: 569–594.

Breuer, K. S. & Landahl, M. T. (1990). The evolution of a localized disturbance in a laminar boundary layer. part i. strong disturbances, *J. Fluid Mech.* 220: 595–621.

Chang, C. L. (2004). Langley stability and transition analysis code (lastrac) version 1.2 user manual, *NASA/TM-2004-213233*.

Enloe, C. L., Mclaughlin, T. E., Vandyken, R. D., Kachner, K. D., Jumper, E. J., C, C. T., Post, M. & Haddad, O. (2004). Mechanisms and responses of a single dielectric barrier plasma actuator: Geometric effects, *AIAA J.* 42: 595–604.

Fischer, P. & Choudhari, M. (2004). Numerical simulation of roughness-induced transient growth in a laminar boundary layer, *AIAA Paper 2004-2539*.

Gaster, M., Grosch, C. E. & Jackson, T. L. (1994). The velocity field created by a shallow bump in a boundary layer, *Phys. Fluids* 6(9): 3079–3085.

Johnson, D. & Rose, W. (1975). Laser velocimeter and hot wire anemometer comparison in a supersonic boundary layer, *AIAA Journal* 13(4): 512–515.

Joslin, R. D. & Grosch, C. E. (1995). Growth characteristics downstream of a shallow bump: Computation and experiment, *Phys. Fluids* 7(12): 3042–3047.

Klebanoff, P. S., Tidstrom, K. D. & Sargent, L. M. (1962). The three-dimensional nature of boundary-layer instability, *J. Fluid Mech.* 12(1): 1–34.

Konrad, W. (1993). Three dimensional supersonic turbulent boundary layer generated by an isentropic compression, *Ph.D dissertation, Princeton University NJ.*

Konrad, W. & Smits, A. (1998). Turbulence measurements in a three-dimensional boundary layer in supersonic flow, *J. Fluid Mech.* 372: 1–23.

Leonov, S., Bityurin, V., Savischenko, N., Yuriev, A. & Gromov, V. (2001). Influence of surface electrical discharge on friction of plate in subsonic and transonic airfoil, *AIAA Paper 2000-0640.*

Muck, K., Spina, E. & Smits, A. (1984). Compilation of turbulence data for an 8 degree compression corner at mach 2.9, *Report MAE-1642, April.*

Rizzetta, D. P. & Visbal, M. R. (2006). Direct numerical simulations of flow past an array of distributed roughness elements, *AIAA Paper 2006-3527.*

Robinson, S. K. (1991). Coherent motions in the turbulent boundary layer, *Ann. Rev. Fluid Mech.* 23: 601–639.

Roe, P. (1981). Approximate riemann solvers, parameter vectors and difference schemes, *J. Computational Physics* 43(2): 357–372.

Roth, J. R., Sherman, D. M. & P, W. S. (2000). Electrohydrodynamic flow control with a glow discharge surface plasma, *AIAA J.* 38(7): 1166–1172.

Samimy, M., Kim, J.-H., Kastner, J., Adamovich, J. & Utkin, Y. (2007). Active control of high-speed and high-reynolds-number jets using plasma actuators, *J. Fluid Mech.* 578: 305–330.

Schmid, P. J. & Henningson, D. S. (2001). *Stability and Transition in Shear Flows*, Springer-Verlag, New York, NY.

Shang, J. S. (2002). Plasma injection for hypersonic blunt body drag reduction, *AIAA J.* 40: 1178–1186.

Shang, J. S., Surzhikov, S. T., Kimmel, R., Gaitonde, D., Menart, J. & Hayes, J. (2005). Mechanisms of plasma actuators for hypersonic flow control, *Progress in Aerospace Sciences*. 41: 642–668.

Smits, A. J. & Dussauge, J.-P. (2006). *Turbulent Shear Layers in Supersonic Flow, second edition*, Springer, New York, PA.

Tumin, A. & Reshotko, E. (2001). Spatial theory of optimal disturbances in boundary layers, *Phys. Fluid* 13(7): 2097–2104.

Tumin, A. & Reshotko, E. (2005). Receptivity of a boundary-layer flow to a three-dimensional hump at finite reynolds numbers, *Phys. Fluid* 17(9): 094101.

Van Leer, B. (1979). Towards the ultimate conservative difference scheme. v. a second order sequel to godunov's method, *J. Computational Physics* 32: 101–136.

White, E. B., Rice, J. M. & Ergin, F. G. (2005). Receptivity of stationary transient disturbances to surface roughness, *Physics of Fluids* 17: 064109.

Worner, A., Rist, U. & Wagner, S. (2003). Humps/steps influence on stability characteristics of two-dimensional laminar boundary layer, *AIAA J.* 41(2): 192–197.

Yan, H. & Gaitonde, D. (2008). Numerical study on effect of a thermal bump in supersonic flow control, *AIAA Paper 2008-3790*.

Yan, H. & Gaitonde, D. (2010). Effect of thermally-induced perturbation in supersonic boundary layer, *Physics of Fluids* 22: 064101.

Yan, H. & Gaitonde, D. (2011). Parametric study of pulsed thermal bumps in supersonic boundary layer, *Shock Waves* 21(5): 411–423.

Yan, H., Gaitonde, D. & Shang, J. (2007). Investigation of localized arc filament plasma actuator in supersonic boundary layer, *AIAA Paper 2007-1234*.

Yan, H., Gaitonde, D. & Shang, J. (2008). The effect of a thermal bump in supersonic flow, *AIAA Paper 2008-1096*.

Yan, H., Knight, D. & Zheltovodov, A. A. (2002). Large eddy simulation of supersonic flat plate boundary layer using miles technique, *J. Fluids Eng.* 124(4): 868–875.

Zaman, K., Samimy, M. & Reeder, M. F. (1994). Control of an axisymmetric jet using vortex generators, *Phys. Fluids* 6(2): 778–793.

Zheltovodov, A. A., Trofimov, V. M., Schülein, E. & Yakovlev, V. N. (1990). An experimental documentation of supersonic turbulent flows in the vicinity of forward- and backward-facing ramps, *Rep No 2030, Institute of Theoretical and Applied Mechanics, USSR Academy of Sciences*.

Permissions

The contributors of this book come from diverse backgrounds, making this book a truly international effort. This book will bring forth new frontiers with its revolutionizing research information and detailed analysis of the nascent developments around the world.

We would like to thank Dr. Mustafa Serdar Genç, for lending his expertise to make the book truly unique. He has played a crucial role in the development of this book. Without his invaluable contribution this book wouldn't have been possible. He has made vital efforts to compile up to date information on the varied aspects of this subject to make this book a valuable addition to the collection of many professionals and students.

This book was conceptualized with the vision of imparting up-to-date information and advanced data in this field. To ensure the same, a matchless editorial board was set up. Every individual on the board went through rigorous rounds of assessment to prove their worth. After which they invested a large part of their time researching and compiling the most relevant data for our readers. Conferences and sessions were held from time to time between the editorial board and the contributing authors to present the data in the most comprehensible form. The editorial team has worked tirelessly to provide valuable and valid information to help people across the globe.

Every chapter published in this book has been scrutinized by our experts. Their significance has been extensively debated. The topics covered herein carry significant findings which will fuel the growth of the discipline. They may even be implemented as practical applications or may be referred to as a beginning point for another development. Chapters in this book were first published by InTech; hereby published with permission under the Creative Commons Attribution License or equivalent.

The editorial board has been involved in producing this book since its inception. They have spent rigorous hours researching and exploring the diverse topics which have resulted in the successful publishing of this book. They have passed on their knowledge of decades through this book. To expedite this challenging task, the publisher supported the team at every step. A small team of assistant editors was also appointed to further simplify the editing procedure and attain best results for the readers.

Our editorial team has been hand-picked from every corner of the world. Their multi-ethnicity adds dynamic inputs to the discussions which result in innovative outcomes. These outcomes are then further discussed with the researchers and contributors who give their valuable feedback and opinion regarding the same. The feedback is then collaborated with the researches and they are edited in a comprehensive manner to aid the understanding of the subject.

Apart from the editorial board, the designing team has also invested a significant amount of their time in understanding the subject and creating the most relevant covers. They scrutinized every image to scout for the most suitable representation of the subject and create an appropriate cover for the book.

The publishing team has been involved in this book since its early stages. They were actively engaged in every process, be it collecting the data, connecting with the contributors or procuring relevant information. The team has been an ardent support to the editorial, designing and production team. Their endless efforts to recruit the best for this project, has resulted in the accomplishment of this book. They are a veteran in the field of academics and their pool of knowledge is as vast as their experience in printing. Their expertise and guidance has proved useful at every step. Their uncompromising quality standards have made this book an exceptional effort. Their encouragement from time to time has been an inspiration for everyone.

The publisher and the editorial board hope that this book will prove to be a valuable piece of knowledge for researchers, students, practitioners and scholars across the globe.

List of Contributors

M. Serdar Genc, H. Hakan Acıkel and M. Tuğrul Akpolat
Wind Engineering and Aerodynamics Research Laboratory, Department of Energy Systems Engineering, Erciyes University, 38039, Kayseri

Ilyas Karasu
Iskenderun Civil Aviation School, Mustafa Kemal University, 31200, Hatay, Turkey

F. R. Menter
ANSYS GmbH, Germany

R. B. Langtry
The Boeing Company, USA

Ünver Kaynak and Samet Çaka Çakmakçıoğlu
TOBB University of Economics and Technology, Ankara, Turkey

Mustafa Serdar Genç
Erciyes University, Kayseri, Turkey

Varun Chitta, Tej P. Dhakal and D. Keith Walters
Mississippi State University, Starkville, MS, USA

Kelly Cohen
University of Cincinnati, Ohio, USA

Selin Aradag
TOBB University of Economics and Technology, Turkey

Stefan Siegel, Jurgen Seidel and Tom McLaughlin
US Air Force Academy, Colorado, USA

Selin Aradag and Akin Paksoy
TOBB University of Economics and Technology, Turkey

Hong Yan
Northwestern Polytechnical University, P.R. China